The Author

Andrew Caine BSc.

Andrew graduated with honours in Marine Biology from the University of North Wales, Bangor.

He has been extensively published in various marine-related magazines for over ten years. During this time, he worked in many areas within the marine industry, concentrating on tropical marine biology.

Andrew also worked in remote Indonesia on coral reef regeneration projects, teaching the local population how to cultivate corals in their natural environment and to help repair damaged coral reefs.

He has now settled down to concentrate on his family and is teaching and inspiring the next generation of scientists.

You may always contact him by email here:

andrewmcaine@gmail.com

Marine Ecology for the Non-Ecologist

Dedications

To Tracey.

Without her support, for me, everything would not be possible.

Marine Ecology for the Non-Ecologist

Published by Andrew Caine

Marine Ecology for the Non-Ecologist

INTRODUCTION

So what exactly is ecology?

In a nutshell, a little known fact is that the fantastic mangrove tree, which thrives in the mangrove estuaries all over the tropics, will grow larger, stronger and quicker if planted on land.

What if we take away the tide? The tide covers the mangrove roots twice every day, so for hours the tree's lower area is underwater - but not just water - salt water. When the tide is out, the tree has to cope with aerial exposure twice daily; plus it also has to deal with another factor, the fresh covering of fine silt from the river. Every day thick mud gets thicker.

Take away all that, and you would have a larger, stronger, healthier mangrove tree growing on land.

However, if a mangrove seed germinates on land and starts to grow, other seedlings from terrestrial trees grow quicker and stronger than our little mangrove. Soon the other trees are taller and are spreading their leaves over the poor mangrove; its fate is sealed. Reduced light means slower growth, a weaker tree, and soon it sadly wilts away and dies. Its competitors' vigorous growth outcompetes it.

It is biologically outcompeted.

Now turn the tables and look at the terrestrial tree shedding its seeds. These seeds fall on the mud of the estuary and the tide comes in, washing them away to rot down - a life never realised!

However, if one lands just where it has a chance, it starts to grow. But the physical aspects of the environment (salty water, very wet mud and new mud every day) cause the tree to grow slower than if it had germinated on land. Here the mangrove can cope with the physical pressures, and soon it's larger, spreading its leaves and shading our doomed seedling, which soon succumbs.

It is outcompeted by the physical and biological pressures of the estuarine environment.

So we have two tree species, both inhabiting and successful in their respective habitats, the shoreline and the tidal range of the estuary. Both inhabit the same ecosystem, albeit in different zones. Both are restricted to their places by either biological pressures and/or physical constraints applied to them as a species.

This is ecology: the understanding of how, as a species, an animal or plant is successful in its particular area or niche within a habitat, and how all these species interact within the more substantial area that is the ecosystem!

It can be complicated and challenging to unravel the strings and knots of understanding, but one thing is certain: ecology is a fascinating and diverse arm of science. Let's take away the scientific terminology, peel back the layers of complication and begin a journey into the beautiful world that is marine ecology.

This book will take the reader on an adventure to discover and understand one of the most dynamic environments on earth. If we look at the planet as a whole, we can identify three significant situations where life exists. These are the earth, air and sea. However, as the place where the land meets the sea, the coastal environment is where animals and plant life have adapted to withstand and succeed in some of the most challenging habitats on earth. We can call this a boundary: where water meets land. While some boundaries are enormous, others are microscopic, such as the water around a grain of sand, where life really does exist.

The boundaries in question are far from static and stable. They experience changes that can be rapid or sometimes very slow, changes that are predictable, or that occur at random. Yet life still thrives and exists in abundance.

It is these coastal realms that we shall visit; we shall see how the environment is defined by the sea and tides. We shall see how the animals have adapted to these conditions and how, as species, they interact with each other to produce the coastline we see today.

However, I would like to take this opportunity to explain one aspect of this book to you before we make a real start. It is what I call 'a random.' This is a piece, added to the end of a chapter, about a random subject with no connection to the section. It is a light-hearted and informative article on a marine topic that is added just for fun, and basically - why not? Enjoy.

Contents

Marine Ecology for the Non-Ecologist

Marine Ecology for the Non-Ecologist

1. ECOLOGY THE BASIC FACTS

The Physical Environment

So, what exactly do we mean by the physical environment? Let's take a piece of shoreline, for example, the rocky shore. We have all been there, playing on the rocks and searching in the pools for life.We all had one thing in common: a desire to locate any form of life that might be hiding there. Remember your excitement! Would you discover more life forms than your family or friends? Would you spot them quicker or find the most unusual or unique animal? You then called out in excitement for everyone to come and look. These are fond memories that we all share.

Now let's do something really horrible. Let's go down to the shoreline at low tide, wade into the water, turn around and look at the rocky shore. There before you is the full expanse of rocky shoreline going right up to the high tide mark. Next, let's have a Harry Potter moment, wave a magic wand, and utter the words 'life extinctus'. In a flash, all life on this area of the shore disappears. By all life, I mean everything from the bacteria covering the surface of the rocks to the algae growing there. All the animals, whether attached to rocks or free moving, have been removed.

What is left?

Most people would say rocks, and they would not be wrong. Looking at the shape of the shore and seeing the lifeless rock pools, you would be forgiven for thinking that was the physical environment. What you would be looking at is only part of the physical environment, known as the topography of the shoreline. That is the shape of the shore.

Is it steep?
Is it flat?
Is it smooth?
Are there crevices and cracks?
Are there boulders?
Is it stable or is it soft and moving?
All this and more must be taken into consideration when understanding the diversity of life that might inhabit this dynamic area.

So far we have only looked at one thing: the shape of the shore. Let's look at the tide. The same pattern happens every day; water covers the rocky shoreline, and then it recedes to expose it again. This happens twice a day, forming a 12-hour cycle. Then over a 14-day period, the height of the tide changes. Each day in the cycle the tide will go higher up and further down the shoreline. The higher the tide, the stronger the water flow, until we reach the highest and strongest tides twice a month – the spring tides.

Depending on where you have set up your house, you are going to be covered by water and then exposed to air twice a day, every day.

We have looked at the shape of the shore and how long it is covered by water and exposed to air. When exposed to air, anything living in that location is also exposed to considerable fluctuations in temperature. Not only that, but they are exposed to these fluctuations very quickly, experiencing rapid changes indeed. We must also consider whether it is raining and these animals are being flooded with fresh water or, indeed, if it is snowing, resulting in the animals being exposed to ice. On the other hand, if it is scorching, the animals and algae are in danger of drying out!

Oh, one other aspect is that water contains oxygen and our animals cannot breathe out of the water. Yet here on the shore they remain exposed for many hours without an abundant oxygen supply!

We have now observed the shape of the shore, its exposure to air, its rapid temperature changes and the fact that oxygen is not always freely available.

To complicate matters, there's more. Is our shoreline exposed to breaking waves with huge swells or will the tide rise and fall gently? Are there fast currents or is there a more relaxed pace of water movement?

When we add all this together and include other factors that are specific to certain environments, like the oxygen content of the sediments, we have what is known as the physical environment. There's a lot to think about.

Marine Ecology for the Non-Ecologist

The Biological Environment

We will now examine how all the animals that are living in a specific area interact with each other. What we will find is that many of the species exist in their own particular areas as we look up the shoreline from the low tide mark. Here we welcome a concept called zonation. Starting from the high tide mark and working down to the low tide mark, we will split all the vertical areas into zones. The splash zone, the high tide zone, the intertidal zone (which is divided into two), and then the low tide zone. Within these zones we will find different species of animals interacting and competing with each other.

How could the conditions created by certain animals and algae allow the development of environments that other species could exploit to their advantage? All this takes place within the biological environment.

Energy Flow through Food Chains and Webs

Without energy there is no life.

You may have seen the films in which a spy, in trouble and about to be caught with no chance of escape, is willing to take their own life. The doomed spy takes a pill out of their pocket, swallows it and lies dead within 30 seconds. Cyanide is the chemical that kills the spy. How cyanide kills is quite simple: it stops the cells of the body from converting glucose to energy.

When this energy supply stops, the body switches off - no energy, no life. It's like a battery-powered toy; when the batteries run out, the toy stops working.

This is what food chains and webs are all about. Most people think it's about understanding which animal consumes another. It is, but more importantly, it's about how the energy contained in a living organism is passed into another body to allow it to grow, repair itself, and reproduce.

We now enter the realm of energy pyramids, food chains, and food webs. We will look at the energy pyramids to find out which animals prey on others. When we explore these interactions, we can ultimately understand how energy passes through the ecosystem to allow the diverse and incredible array of animals to exist in the populations we see today.

First, we will observe how energy passes up through the ecosystem.

The primary producers are always at the bottom of the pyramid, and 99% of the time, they will be plants and algae. The other 1% are bacteria. What these fantastic life forms do is to allow all life on earth to exist. Take away the primary producers, and we remove the first step in the food chain. As a result the food chain will collapse, and everything will die.

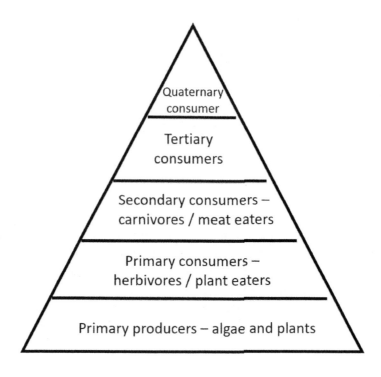

Doodle 1. Biology separates different species of animals into groups based on how they obtain their energy.

Through a series of complex biological reactions within the plants' cells, they trap the sun's energy, namely light, and then convert it to produce glucose. This glucose, or should I say sugar, converts to energy which allows the plant or algae to grow. Grazers or filter feeders then consume this growth and the energy passes up the food chain. Plants, algae and some bacteria are the primary producers because, quite simply, they are first in the pyramid and produce their own food.

We then move up to the primary consumers, for the first time seeing complex animals, which are always herbivores. What this means is that they feed only on plants. We notice that the pyramid is now getting narrower as there is a smaller amount of biomass than within the primary producers. What do we mean by biomass? Put simply, if I were to place you into a blender, turn it on, mash you up into a pulp and then pour you out into a big bucket, I could then measure what was in the bucket and that would be your biomass.

Next, we move up the pyramid into the third layer and find our secondary consumer. Here we meet the first predators: the hunters, or animals that obtain their energy by eating other animals. They are the primary true carnivores.

In areas where we have a huge food web, we will find that these predators are quite small; however, in areas with short food webs, these can be quite large animals indeed. Again, as we move up the pyramid, we see it is getting thinner. Not only can we consider, more importantly, that the total biomass is smaller, but we can also look at the species number. In the marine environment when we look at the herbivores, the primary consumers, we will always observe a vast number of species. When we look at secondary consumers we will see there is a significant reduction in species number!

As we go higher, we meet our tertiary consumer, a large complex carnivore indeed, whose only goal in life is to eat and reproduce. Then up we go, even higher, to find the quaternary consumer. Again, we see the number of species, and indeed the total biomass, is reduced. Here we find the apex predator, the animal which no other feeds upon when it is alive and swimming.

Doodle 2. A generalised adaptation of the pyramid showing examples of animals found within each layer.

So we have seen that ultimately all the energy that flows through animals in the world's oceans starts as light rays from the sun. We have seen how the number of species reduces as we move up the pyramid of life.

The question is: why does the number of species reduce as energy is transferred from one level to another? If we lived in a closed system where no energy was lost or removed, then this would not be the case. 100% of the energy transferred would stay in the animal until it is eaten and all that energy would be passed up a level too.

The law of conservation of energy states that energy cannot be created or destroyed but only changed in form. It is to this law that we must look when we see what happens to energy as we move up the pyramid.

Enter the 10% rule of thumb for energy transfer as we move up the pyramid. Simply put, if a fish was swimming around and it came to a box, which contained 100% energy, and the fish consumed it, then 90% of the energy would be used or wasted. The fish itself would then be consumed. However, 90% of the box of energy would be absent, as only 10% of the total energy is transferred up into the next layer.

Now that is quite frightening when we consider that, of the 100% of energy that is received by the plants and algae, only 10% will go to the next level. When the herbivore is consumed, only 1% of the original sun's energy is transferred to the next level. Going right up to the quaternary level, where we find the apex predators, we see that only 0.01% of the sun's energy is left. This means that this tiny amount of energy is all that remains from the original 100%. This is the reason most pyramids will end at the quaternary level because, to put it quite simply, the energy has run out.

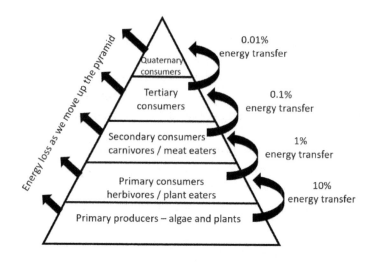

Doodle 3. Energy transferred up the layers of the pyramid.

Let us return to our box of energy that the fish found and consumed and pose the question: where did all the energy go?

There are several ways the energy is lost. The first is, basically, not everything that is eaten is digested. This undigested food represents a large amount of energy that is unavailable to the animal.

This energy is lost as poo. However, the energy is not lost from the system as a whole. Let us now look at a group of organisms known as the decomposers. Anything that feeds upon waste is grouped into this bundle of species. Crabs, worms and shrimps are the large decomposers, and if we take a closer look through a microscope, we discover the true realm of decomposers: the world of bacteria. This greatly important group of organisms feeds upon waste, breaks it down to the molecular level, and the end product is then absorbed by the primary producers and passed back up into the pyramid of life.

Not only this! Our minibeasts, the decomposers, also perform another vital function in the decomposition of dead bodies. You would think if anything died in the marine environment, it would be scavenged very quickly, which is typically the case. However, at any one time, a large amount of biomass will be floating around or buried in the sand, all its energy trapped. The same beasts that eat the poo will also feast upon the dead animal remains and pass that energy back up the pyramid.

So what happens to the rest of the energy that is lost? Well, quite frankly, it is used to keep the predator alive when it eats its prey. It is like eating a battery full of energy; the energy contained in that battery will be consumed to keep the animal working, but then the battery needs to be recharged by the next meal. The animal needs to reproduce; or it could be injured, and energy will be used to repair its body. A large chunk is also lost as heat. So when we consider all of the above, we can see why only 10 % is transferred up to each level in the pyramid.

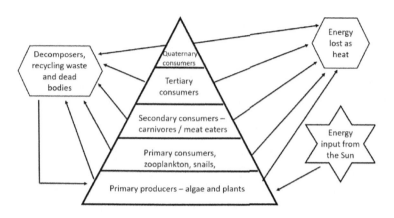

Doodle 4. Energy loss and input in the pyramid of life.

Now we can leave the pyramids behind and look at precisely what eats what, entering the domain of food chains and webs. Food chains are the most straightforward type of predator-prey relationships we find when examining the complex ecosystems of coastal ecology. They allow us to look at which animals eat which within a habitat. All food chains start with the primary producers and then move up to include the predators and prey species.

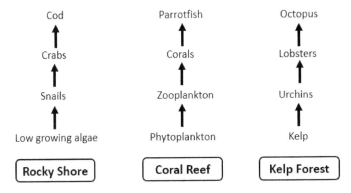

Food chains are the easiest predator – prey relationship to study. However in reality they do not exist, as more than one species eats another.

Doodle 5. Three short food chains within different marine environments.

Food chains are a means for us to study predator-prey relationships in simple form. However, when looking at whole ecosystems in reality, we find food chains rarely exist. Here we move into the realm of the food web.

When we look at a food web, we usually do so in a simplified form. Very rarely, unless we are at the tertiary and quaternary level, do we name individual species in the food web. What we do is generalise by saying shrimps, crabs or worms. We have to do this because if we named every species that existed in a habitat, either the paper that we were writing on would be way, way, way, way, way too small or the print font that we used would have to be so little that we would need a microscope to read the letters. Basically, there are too many species present in their habitats to be individually

23

named in order to produce an accurate and complete food web for the area.

Even when we produce simple food webs, we find they are quite complicated. It takes time to follow exactly what they are showing. The lines in a food web connect a prey animal with its predator and, quite often, one prey item is fed upon by a few predators rather than just one.

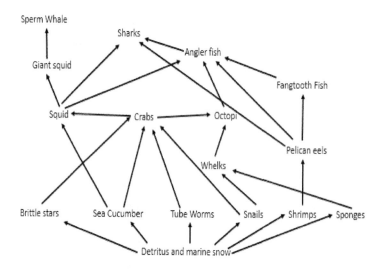

Doodle 6. A simple food web for the deep-sea environment.

In the deep sea, it is clear that the primary producers are the decomposers, for no light has ever penetrated down to this level since the oceans first filled with water. The energy input for this food web comes from dead animals sinking from the upper regions, but most importantly, a constant but slow falling of dead plankton. There is so much of this biomass that when you look at it through the window of a deep-sea submersible, it looks like snow is falling from the skies; hence its name: marine snow.

We can see here that the primary consumers are all decomposers, feeding on this supply of dead organic energy. Then, as we move up the food web, we notice there is a massive drop in species number just like in our pyramid.

We can observe that if the pelican eel died out due to disease, we would also lose the fangtooth fish. The angler fish and deep-sea sharks would survive as they have other species of prey to feast upon. However, it would be bad news for their prey items, which would be hunted more intensely to compensate for the loss of the fangtooth fish.

This is the arena of predator-prey relationships. It really does not make a difference which relationships we look at: it could be lions and zebras, foxes and rabbits, or crown-of-thorns starfish and corals - they all follow roughly the same pattern.

The predator population has a specific size, and each member of a community has to acquire food in the form of prey. This food not only keeps the predator alive and healthy (most importantly), but it also provides food for the predators' offspring, which are the next generation. At this point, the predator's population is high, so there are enormous pressures put on the prey population to supply the predators and their offspring with all their nutritional needs. Soon the prey population starts to fall, as not only are the adult prey removed, but so are their offspring - either falling foul of our predator or starving to death as their parents have been eaten. With the loss of prey numbers, the predators themselves start to suffer. They can't find enough food, their offspring succumb to starvation, and very few will reach adulthood. Not only that, but the parents themselves are weakened through lack of food and thus either die of starvation or disease. Now the predator numbers are falling, and with fewer predators around the prey, parents are not only staying alive but they are raising more offspring to reach adulthood successfully. With this increase in prey, we find that predators are acquiring more food, and their populations are now rising. Get the picture? It's not the circle of life you're used to, but it is indeed a circle of life.

Earlier we briefly touched on what happens when prey items are removed from a habitat. I think it's essential to look at what has happened in the past, and what can happen to an ecosystem, when humans overfish a single predator species that was also the prey of another animal.

Here we look to the kelp forest. Giant kelp can grow over 30m tall in the ocean; indeed a magnificent underwater forest. Inhabiting this unique habitat are hundreds of species of animals, as you would expect.; yet one was of high commercial value. In other words, this species was worth a lot of money to the fishermen, and more so, the restaurants. This species is the lobster.

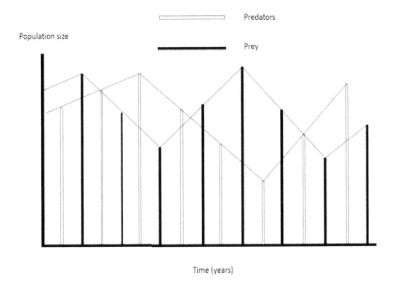

Doodle 7. Typical predator-prey relationships over many, many years. Notice how the populations of both predator and prey will rise and fall over time.

Now, the bad news for the lobster was that people in the restaurants could not get enough of it. When there is low supply of a product (the lobster) then the price goes up. Soon the fishermen gained very high prices for their lobsters - happy days for the fishermen and restaurants, but not-so-sunny days for the lobsters.

Soon the lobsters were being rapidly overfished. Their population was being depleted; it was getting smaller and smaller and smaller. What did this mean for the price of the lobster? Well, it got higher and higher and higher as there were very few around. Eventually the lobster population was utterly decimated.

Now, this was terrible news for the octopus, which liked to feed on the lobster, so its population started to decrease. However, it would also eat crabs and other crustaceans, so their population dropped, and more pressure was placed on the crabs to survive.

With the removal of the main predator, the lobster, the urchin population exploded and it did so rapidly. What a time to be an urchin -no predators- happy days indeed! The urchin is a grazer, a primary consumer or herbivore. With a vast increase in population the urchins had to find an additional food source and they did: the holdfast of the kelp. The holdfast is the part of the kelp that anchors this massive algae to the rock. It keeps it in place and holds it down, attaching it securely and firmly, like roots hold the oak tree in place. Soon the urchins were stripping the kelp of their holdfasts, and the kelp were simply floating away. With the kelp floating away and no algae to graze on, the urchins found themselves in trouble too and their population also crashed.

After the grazers and the kelp were removed from the habitat, new forms of algae were able to take hold. A considerable number of species that inhabited the 30 m kelp forest had nowhere to live; their cover was destroyed and soon they were gobbled up by many happy predators.

With the new forms of algae inhabiting the area, came new species of herbivores, which were followed by a new species of predator. In a short time, the habitat's whole food web changed. In fact, the entire habitat itself had altered from kelp forest to rocky reef, all because people were willing to pay high prices for lobster in a restaurant. This is a different form of ecology, where unnatural factors disrupted the predator and prey relationships and the natural progression of the species, resulting in a completely new habitat being formed. And guess what! There are no lobsters.

Here we enter the dominion of what is known as the keystone species. This species is hugely important to the makeup of its habitat. If something happens to the keystone species then the whole habitat structure will change. In our example, the lobster was the keystone species. Although not mentioned by name, there were many more animal species in that habitat; yet the removal of one of those would have had minimal effect on the whole habitat structure due to the diverse food web. Look at a food web and think about it! The arrows will shift and move around, but the web will only alter a small amount, going back to its original shape over time. However, this is not the case if you remove a keystone species.

Marine Ecology for the Non-Ecologist

Food chains are the easiest predator – prey relationship to study.
Look what happened to the kelp forest in US waters when man overfished the lobster.

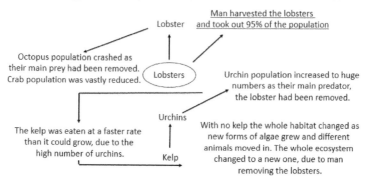

Doodle 8. How the removal of lobsters disrupted and changed the original habitat to a completely new one.

Take the Antarctic habitat, where we have a species of shrimp called krill. These are the keystone species in the area. Remove the krill, and the many species of fish dependent on them would also decrease to a critical level; the populations of the seals and birds, which feed on these fish, would deplete and the magnificent great whales would also be in trouble. The whole food web would collapse, and thus, a change in the ecology of the area would occur just by removing a shrimp!

It is the keystone species that holds the whole ecosystem in place, just like a keystone in an architectural building supports the entire structure. Remove the keystone and it all falls.

Random 1.

The Immortal Jellyfish

I was born in the Mediterranean Sea, just off the coastline that you humans call Malta. It was the year of our lord 1534, and since that time I have witnessed many wonders. I have also observed many disasters; I have seen my brothers and sisters die, yet some are still around. My great, great grandad, the oldest in our bloodline, is still going strong - still hunting down prey. He was born in the year 1129 and doesn't look a day older than two years old. We know how to keep our skins silky soft!

We love the Mediterranean; it's not too cold, not too hot - just right. So much so that our species did not want to leave this area; why move? Abundant food, gorgeous lady friends, and we can drink from the fountain of youth. Why move indeed? As a species we existed for thousands of years here until recently, when an enormous metal whale sucked in hundreds of our family members, never to be seen again (or so we thought). Eighteen years later, one made it back. Apparently it was something called a ship that sucked in water to stabilise itself. When it reached a place called Japan it spat out my friends. They soon made a home and we now exist all around the world, thanks to the metal whale called a ship.

Marine Ecology for the Non-Ecologist

Anyway I am waffling… Yes, I am a bit of a chatterbox; I do like the sound of my own voice – deal with it!

So who am I? You lot, the humans, call me the immortal jellyfish; for that I am. I am old as I have said but in our terms, a mere baby. I am immortal: I cheat death, I regrow, and only a predator can stop me. Get the picture?

Now you lot, you need to understand death. Understand the process of old age and natural demise - the scientific term is phenoptosis. In this process, an organism's genes include switches that under certain circumstances cause its cells to die, and thus, the organism degenerates and dies. Timed cell death.

To understand me, you also need to know how we jellyfish reproduce. In a nutshell: (as I don't think you are very intelligent) we start off as microscopic plankton, then settle out, attach to rocks and transform into anemone-like forms. We develop and then spit out from ten to over fifty little mini-jellyfish - the young adults.

In fact, I am getting a little tired. I had an accident last week and lost a tentacle. I was then infected by some bacteria - they are out of control, little sods. This has happened to me countless times and I am bloody-well fed up with those microscopic bugs. However, I can feel phenoptosis about to kick in; time for a change.

Looking down I can see a nice rocky ledge and just in time too. OK that's it; that's my new home. I'll go and attach. You know, I love this part of my life. I have just stuck myself to the rock and now it's time for the fountain of youth to be sipped.

I have never seen one, but could this be true? Could a thing called a butterfly start as a caterpillar - a crawling thing - then go to sleep in a cradle and awake as a different animal - a flying beast? Is it true? If so, it is doing things the wrong way around. Think about it: it is changing from a young beast into an older beast, hurtling through life getting older each day; phenoptosis here we come.

Well, we do it the other way around. I am an adult; I am tired, I am sick, I am dying, and I am now stuck to a rock. Here's the trick: my body's cells undergo transdifferentiation, just like that caterpillar. Muscle cells become eggs, heart cells become stingers, gut cells become muscle cells - and you get the picture. I change on a cellular basis. I renew, but unlike the caterpillar, I switch direction and become that juvenile polyp - the baby. My adult form returns to its beginning. I am reborn. I have drunk from the fountain of youth; yes, it really exists. Well, it does if you are me!

A few weeks later, and I am off again, a young beast of only 486 years old and getting younger by the year. The only thing that can stop me is a predator; which reminds me, I have been waffling too much - and to lose concentration in my world is to lose your life, so goodbye.

2. THE PHYSICAL ASPECTS THAT SHAPE THE COASTAL ENVIRONMENT

One thing is never more important than another when considering all we need to look at in the physical environment. Instead, there are many critical physical conditions that combine to create the habitat that we see.

The first of these is water movement, and of the many types of water movement that shape our coast, we shall examine the four major players. They are waves, tides, currents and aerial transport of spray. Having said that, we have one more baby at play here: the introduction and loss of freshwater to and from the saltwater environment. We will meet that one later.

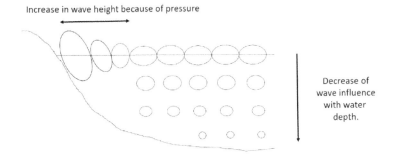

Doodle 9. How waves moving towards the shoreline alter, both as they approach the shore and with depth.

Now waves are funny things. You would think, when watching waves, that the water is either moving towards you or away from you. However, what is really happening is that the water is moving up and down while the energy is moving towards you or away from you. Hard to visualise isn't it? Yet the water particles are transferring the energy in the direction of the wave whilst moving vertically up and down.

The waves we look at when considering coastal ecology are the ones that are crashing or merely lapping on the shore. 99 % of the time, these are wind-generated waves. Go down to the shoreline on a hot windless summer day and the water will be as smooth as a baby's bum. The next day we could go down to the same place in the middle of a gale, and we would have 3m-high waves crashing into the shoreline.

When we look at how these waves are made, we find there are three principal factors interacting to produce waves that are either a surfer's paradise or a skimming stone's delight. These are the wind's speed, its duration and the uninterrupted distance across the water that the wind is acting on - called the fetch.

Standing in Ireland and looking to America, we will have 5600 km of Atlantic Ocean for the wind to act on. If we look to the west, there are only about 400 km until we hit the UK, so we can see here that fetch is indeed critical.

Let's start a few miles from the coast, with a gentle wind blowing. Here we have a nice swell: a smooth up and down movement. As we move closer towards the land we pass over a rocky reef, where the water shallows and the rocks cause an obstruction to the water movement. It's like holding your hand up against the wind in a severe gale; the air movement is restricted around your hand, which disturbs the air flow as it moves past. This is what happens to the waves. They scrunch up, relax then scrunch up again, which results in the gentle swell being completely transformed into a very choppy and unpredictable surface. As we move over the reef and pass it, we return to a gentle smooth swell.

Moving towards the shore, the water depth will gradually decrease. So this effect is seen again; the waves become increasingly distorted due to the energy building up in the water until they crash into the base of the cliffs. Now, if we are on a sandy beach, the depth may decrease in a smoother and more uniform manner, and the water surface will be far less chaotic than that of the rocky shore. In both instances, the higher the wind speed, the greater the effect.

There is one really nasty wave that is not wind-formed, but is created by a sudden release of energy from the earth's crust (such as an earthquake on the seafloor), and that is a tsunami or tidal wave. Thankfully it is a very special, random and rare event.

So how are tsunamis created? Well, we have a disturbance releasing energy into the sea and this sudden injection of power creates an exceptional wave moving out in a circular manner from the centre point. When we look at the tops of waves, called crests, and measure the distance between two crests, this is called the wavelength. Now, the tsunami has an enormous wavelength. It would be like looking at a completely flat sea, where the distance between the two crests could be as much as 200km. Yet the wave height, or amplitude, would only be as much as one metre in height. This would give us a very, very, very long and shallow wave. Another unique feature is the speed of the energy contained in the moving wave, which could be up to 600 km/h.

This is not a problem when we're out in the ocean. However, as the wave approaches land and the sea shallows, this causes the wave to crunch up. As it crunches up, it increases in height. Now, this baby has 200 km of shallow water moving at incredible speed and suddenly meeting land. The wave height may build up to 10 m or so, forming what is simply an extremely thick and very high wall of water. This will crash into the coast and keep moving inland until the energy has left it, causing immense devastation and loss of life. The worst on record was the 2004 Boxing Day Tsunami in Indonesia and Thailand that caused over 250,000 deaths, where whole villages, communities and holiday resorts were wiped out along with the people in them.

From the human viewpoint, this was a disaster of unprecedented proportions. However, when we look at it from an ecological perspective, it would be called a biological disturbance. It was a catastrophe for humans because of the massive loss of life and buildings, but when we look at how animals colonise different habitats, we find that it wasn't long before new and different species were moving into the area that had been cleared by the tidal wave. Life moved on. What happened before any point in time was just a disturbance in the natural process that is ecology.

When the waves crash at high tide they form a spray, which is blown inland by the wind. 'Big deal', I can hear you say. Well, this spray is not just water but salt water, and this salt water is poured over the area twice a day, every day. Now, the build-up of salt in the soil or over the rocks stops many plants from growing. The habitat is influenced by the sea. This is the limit of the marine environment's influence, and beyond this point is where the exact terrestrial habitats begin. It is the splash zone and marks the end of the marine environment.

Tides are funny things. Indeed, you would be forgiven, when standing on a beach and looking at the water as it races to your feet, for thinking that the tide was coming in. The tide is coming in, and soon you will be up to your knees in water and have to move back towards the shore. Is it the water that is moving up, or is it the earth that is slowly spinning in its 24-hour cycle? The latter is true. The water does not move up and down. It is the earth that is rotating, and the moon's gravity is holding the water in place.

To explain this, we have to take the earth and smooth it off into a perfect ball shape. All the mountains are removed and all landmasses are thinned into a perfect sphere. What we have is a ball of rock covered by the water that is the world's oceans. Here, the spinning earth's gravity tries to pull the water towards its centre while at the same time the moon's gravity is trying to draw water away from the planet. It's like a tug of war between two rocks - only, the more massive rock will always win. However, as the moon moves around the earth, we will see a bulge in the ocean. Where there is more water, the tide is in, and where there is less water, the tide is out.

We have seen previously how the tides differ in a 14-day cycle resulting in the most significant and lowest tides every 14 days: the spring tides. However, if we consider the tides on a global basis, we find that they have less influence in the tropics, where the tidal range is less than 1 m. This increases as we move further away from the tropics, where we can have extreme spring tides ranging up to 17 m.

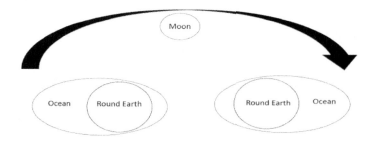

Doodle 10. The ocean bulge if we had a perfectly round earth, with the moon's gravity pulling on the ocean.

Another hugely important factor is the currents existing in the bodies of water that we find within the marine environment. Important factors creating these currents are the ebb and flood of the tide, but most important is the horizontal drag produced by the wind on the surface water.

Now, ocean water has very distinct characteristics that differ from one area to another. These are temperature, salinity and chemical makeup. So we have a situation where two bodies of water with different characteristics meet; quite simply: they will not mix. Think of it as a long road where you have traffic moving in one direction and, on the opposite side, all the traffic is moving in a different direction. In between these you have a safety barrier. The barrier is where the two bodies of water that are moving in different directions meet, but because of their chemical makeup and temperature, they do not mix. Here we have two different currents side by side, creating completely different environments for animals to exploit or avoid. An excellent example of this is the Gulf Stream, which originates around Mexico and flows northwards. As it is warm water, it will float on the surface of the Atlantic Ocean and flow up to the north of Scotland, where it then cools and falls back into the depths. This warm water influences the climate around the entire UK and keeps it relatively warm. If you look to the west of Scotland you will find Canada, where winter temperatures can drop as low as -40° C, yet in Scotland -15° C is about as bad as it gets, kept warm by the Gulf Stream.

Another hugely important current is the upwelling. This occurs when the warm surface water in the tropics is pushed away from the land by the wind. The surface water is generally nutrient deficient and supports a low level of life. As this water moves away from the land, it drags cold, nutrient-rich water up from the deep sea. This cools the area and allows a high diversity of life to exist.

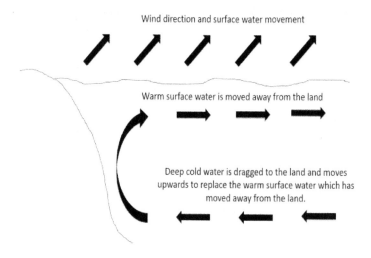

Doodle 11. The wind-powered upwelling of cold, deep nutrient-rich water.

Currents around the coastal area vary, and if you don't know the region, they can be very unpredictable and dangerous. One moment there is little water movement, then within 2 or 3 minutes, the water can be moving at high speed. It's like turning a switch from low-velocity water to high within a very short time. It can be very dangerous, so it's always a good idea to get local advice or even watch from the shore over a tidal period before you go wading.

As you would imagine, and as we will see later, these currents significantly affect the diversity of life that is found within an area; high flow and animals could be washed away!

Temperature is critical to life: too hot, too cold (even by just 1° C) can lead to an untimely death. However, rapid changes in temperature can have the same effect on the little beasts inhabiting our coasts. So a stable temperature is not just one that remains the same but also one that changes over a long period of time, allowing the animals to adjust. Temperature in the tropics and polar regions varies slightly throughout the year, whereas the water temperature in the temperate regions experiences fluctuations from 3° C to 20° C over the seasons.

With temperature, we discover the phenomenon called the thermocline. You might have seen the film *Meg*, a Hollywood blockbuster all about an extinct shark - the megalodon - being found alive and well. The sharks had existed for millions of years under a thermocline in the deep sea until a couple of them broke through this barrier to wreak havoc. So, the thermocline is a physical barrier between two water masses deep within the sea; a barrier created by different temperatures. In temperate waters there is a deep-water body of 2° C and an upper layer of up to 20° C.

We have thermoclines that exist all of the time, and we have seasonal ones. The latter are created by the warmer waters of spring and summer that break down in winter due to the mixing effect of increased wave influence. This breakdown is significant, as the surface waters are now nutrient deficient - all the goodies taken out by the summer growth of plankton. With the breakdown, nutrients locked up in the deep sea are mixed into the surface waters ready for the plankton bloom in spring. It allows life to thrive in the summer months.

Salinity or salt content is the last of our physical aspects to consider, and salt water around the coast is generally between 33 and 34 parts per 1000 - typically written as 34°/oo. This will usually remain constant around the shores of the world. However, variations do exist where we get more evaporation than rain, such as in the Red Sea or, more importantly for this book, in areas of isolated water such as rock pools. We will look at this later when we examine the problems it causes and how animals must overcome them.

Random 2.

The Blue Dragon

I am a member of the oceanic life existing on this wonderful planet, and there is a single thought that amazes all organisms inhabiting this aquatic world: how do you humans have such a high regard for your own intelligence, yet you know so little about us?! From the microscopic to the friendly giant squid 2500 metres down, we all think the same thing - you've got a lot to learn.

I am blue. Well, part of me is. However, I am not 20 metres long, I do not have huge powerful wings, I exist in the sea so do not soar into the skies above, and I most certainly do not breathe fire. All this, and you lot have given me the name: blue dragon - speechless. My real name is *Glaucus atlanticus* and let me introduce myself (my real self) as I am quite an interesting beast indeed. Remember me, if you ever get the lucky opportunity to see me!

I am a member of the molluscs, the second largest animal grouping, with over 85,000 cousins ranging from the giant squid to my microscopic mates. My nearest relatives are the nudibranchs that you like to call the sea snails. However, I am quite different. Unique you could say, a terror too. Delicate yet extremely hardy; I am a force to be reckoned with.

I am a lazy sod who likes to sunbathe, so I try not to swim. Why bother if you have an air sac in your gut to help you float - why waste energy? Floating is the norm, but I am a good paddler (more about that later). My floating habits keep me at the mercy of the ocean currents, which means that I am found all around the globe, inhabiting temperate and tropical waters alike; everywhere except the polar seas - cold kills!

So there I am floating around - but I float on my back, upside down. (Well, I like to see what's in the sea around me, not what's in the sky!) My beautifully belly is the most iridescent blue, so birds cannot see me, and my back is white, avoiding being silhouetted against the water surface when predators look up from below. Crafty, isn't it?

I rest with my arms stretching out and splitting into many folds, like branches on a tree, each poised to deliver a killing punch to my food source. Well, a blokes got to eat hasn't he? I said you humans don't know much; here's another thing that's got you baffled. I sting my prey to death, but I don't grow any stinging cells. Instead, I rob them from the very thing I eat – nice.

Portuguese man-o'-war: not a jellyfish but a colony of individual animals suspended beneath the float, is my main course tonight. There it is, drifting by. I see it, and it is doomed. It can only float, but I can paddle, and soon it is in my deadly grasp; an army of sting harpoons erupt from my arms, piercing my prey's body. Soon I start to eat. Now, the cells I am talking about contain a harpoon and venom; each cell has a hair on it that when pressed, causes the harpoon, kept under high pressure within the cell, to fire.

What I do next is top secret; I eat the stinging cells, but they do not fire! They enter my gut with all its digestive juices and get pounded around as I digest my meal, but they do not fire. They then leave my gut and I transport them to my arms. I locate these cells at the surface, and again they do not fire. Now they are ready to be used the next time I meet my prey. I ever-so-lightly brush against my prey, and all hell erupts! Work that one out if you can...

I am a perfect predator: my prey cannot escape, its body parts are utilised for my defence and acquiring a meal, and I have perfect colouration. I must add that I am a very beautiful, yet deadly beast indeed. The only thing that really upsets me is that I am only about 4 cm long!

3. THE ROCKY SHORE

Here we have the most diverse and most studied of all the coastal marine environments. Depending on location, the rocky shore could be sheer cliffs rising hundreds of metres out of the sea, smooth slopes gently disappearing under the water, ledges split by deep crevices, boulder beaches or tidal rapids. Many of these can be seen in just one location interspersed by an array of rock pools. Are the rocks we are standing on very hard? Or are they quite soft, flaky and subject to erosion? When you compare this to the sandy beach, we have an extremely varied and dynamic environment.

However, when we find ourselves at the high tide mark, looking down the shoreline, we observe one physical aspect governing the diversity of life that is more important than any other. Are we looking at a high or low energy shoreline? The two things to consider here are the waves and the speed of the current.

The higher energy environment will either have massive waves crashing over the rocks or fast-moving water continually acting on their surface. In our latter example: imagine your hand is a nice big clump of seaweed, and the back of your hand is forced continuously against a rough piece of sandpaper (rocks) by the wave action. Quite soon you will find that there's nothing left of the back of your hand, and that is precisely what happens to any algae trying to grow on a wave-exposed rocky shore.

So when we don't see algae, we know we've got waves. The fast-moving currents also cause problems for animals, which have to hang on to avoid being swept away. Here algae have no problems growing; they just bend with the currents, like trees in the wind, never touching the rock surface.

Lower energy environments allow a more diverse array of species to exist, merely because of their relaxed physical pressure. So when we look at life on the rocky coast, we will split this into two sections: the high energy and low energy environments. Before we do that, let's examine some of the physical pressures applied to our beasts that make living here difficult. We will discover how they have adapted to and not only survive, but thrive in these environments.

Temperature, depending on where you are on the rocky shore, can have a substantial or minimal effect on your body. We find that we have different adaptations ranging from the tropics to the polar seas. When we say temperature, we really mean heat shock: suddenly you are exposed to air, and the protection of the water has left you at the mercy of the sun or the frost for a brief moment in time.

I can remember many years ago, walking along the rocky shore in Wales in early spring. Something was troubling me; the life was gone - no barnacles, mussels or limpets. Then I remembered that we had been subjected to the most severe winter in living memory. The tide had gone out at night, and suddenly the animals were exposed to extreme temperatures; so quite frankly, everyone had frozen to death. Returning the year after, the shoreline was full of life as the planktonic larvae had settled and repopulated the shore. That severe winter was another example of a biological disturbance.

We find that in the tropics, many of the shellfish or molluscs have elaborately sculptured shells, increasing their surface area and acting as radiators to remove the heat. We also see that many have colouration that reflects the sun's radiation, keeping them cool. In the temperate zones, we find snail species whose cells can withstand temperatures exceeding the higher and lower temperature ranges found in the area. All of the highly mobile species either retreat to rock pools, hide under algae or stones, or run down the shore as the tide recedes and are never exposed. What we do find is that different species have different tolerances, and those that can withstand higher degrees of heat shock are located higher on the shore.

Along with temperature, comes another physical pressure called desiccation or merely, drying out. The longer you're exposed to the sun, the more at risk you are of drying out. We find different species of barnacles inhabiting particular areas purely because of the way their interlocking plates fit together. A tighter fit means less water loss; thus, these beasts are quite happy on the high shoreline. Lower down, we find a more aggressively growing species but with looser fitting plates. If our high tide species grew here, it would soon be forced off the rock by the other faster-growing and more aggressive species. However, if the fast-growing species tried to grow in the upper regions, only a few low tides on a couple of hot days would see them dry out and die.

Sunlight or light intensity can also cause problems for our primary producers, the algae. Again, we find the most significant variation in the earth's temperate zones. Quite commonly on the rocky shore, you will notice the green low-growing algae high on the coast, growing well and vigorously in the spring months. This causes many problems for us humans, who like to go rambling or fishing, as they are quite slippery underfoot. These algae have resulted in many broken bones. As we move into late spring and summer the light intensity increases, and more importantly, the UV content of sunlight is higher. We find the algae cannot cope with this UV intensity so are short-lived. This is often termed by fishermen as 'the algae have burned off.'

Let's look at some feeding strategies existing within our animal population. The most common life form we see on the rocky shore is that of the sessile animal. These animals settle out from their plankton stage, attach themselves to rock and then cannot move. They are filter feeders that obtain nutrients by capturing phytoplankton, zooplankton, or any organic material that is floating around in the sea. They can do this by sucking sea water into their bodies and collecting the food it contains or by casting elaborate nets and pulling in their harvests.

We then move on to the mobile animals, starting with our herbivores. Each species has evolved a different method of removing algae from the rocks or extracting layers of cells from the central part of the algae. Examples are urchins, chitons and the mobile snails.

Next we find our predators: crabs, whelks, starfish, birds, and fish themselves. Some species are highly specialized and feed on one prey item only. Most, however, are able to utilize many species to fulfil their requirements. Yet most of these animals can only capture and consume prey of a specific size. Here we encounter the phenomenon called the refuge. If a prey item is able to exceed a particular size, it will avoid being eaten. By achieving something that is called the size refuge, its size will save it.

Now, when we look at predator activity, we find they are only able to hunt and capture food when it is covered with water. So if you live in the high tide regions, you are more exposed to air but have less time exposed to predators. Therefore, as a species, despite being forced into an area that is less productive for you, you're going to stay alive. This is called the spatial refuge: the locality where you will not be predated.

Are we starting to get the picture? Predators cannot reach the prey, some animals do not grow as vigorously as others, some can tolerate heat shock yet others cannot, and certain species are resistant to desiccation whilst others will dry out quickly. And we've only just begun!

When we consider the above, we will see, as we move from the low tide mark up to the high tide and splash zones, a distinct species distribution. Here we unpack the concept of zonation.

Scientific literature can often cite over seven different zones for the rocky shore, not including any micro-habitats existing in the area. It gets complicated - really complicated, so for our purposes we will look at zonation in a more simplified form.

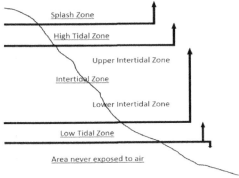

Doodle 12. Our simplified version of the zonation of the rocky intertidal shore.

If we start at the bottom of the shore looking vertically upwards, our first zone is the low tide zone: the area that is exposed from the neap to the spring low tide (at its lowest point.) We then move to the intertidal zone, which is split into two: the lower and upper intertidal zones. Then we see the high tidal area, again reaching its peak at the spring tide. Last is the splash zone, where we have the influence of saltwater spray. This represents the boundary of the marine environment.

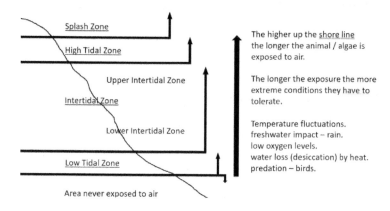

Doodle 13. How exposure changes as we move up the shore, and the implications it has for life there.

So we have our zonation, but what will live and thrive within each zone? Also, we must consider: are we in a waveswept or relatively calm environment? An enormous diversity of life is found within the lower energy environments as illustrated by two different doodles showing the zonation of the animals and algae.

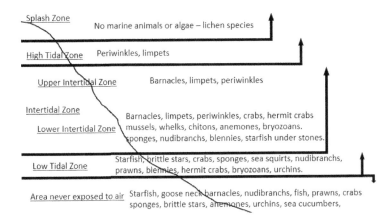

Splash Zone — No marine animals or algae – lichen species

High Tidal Zone — Periwinkles, limpets

Upper Intertidal Zone — Barnacles, limpets, periwinkles

Intertidal Zone

Lower Intertidal Zone — Barnacles, limpets, periwinkles, crabs, hermit crabs mussels, whelks, chitons, anemones, bryozoans. sponges, nudibranchs, blennies, starfish under stones.

Low Tidal Zone — Starfish, brittle stars, crabs, sponges, sea squirts, nudibranchs, prawns, blennies, hermit crabs, bryozoans, urchins.

Area never exposed to air — Starfish, goose neck barnacles, nudibranchs, fish, prawns, crabs sponges, brittle stars, anemones, urchins, sea cucumbers,

Doodle 14. A generalised description of the animal species inhabiting the rocky shore where wave influence is low. Moving down, there will be more species of sponges in the lower tidal zone than the lower intertidal zone. The number of species will increase as we move closer to the ocean.

Another type of zonation in the low exposed area is that of the vastly significant algal growth that occurs there, with seaweeds being our primary producers. Again, we have a wide range of factors that allow or restrict the algal colonisation of a particular zone. Aerial exposure, desiccation, light intensity, predation and abrasion all play their part.

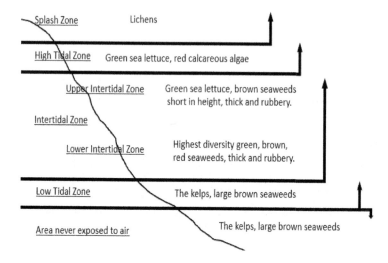

Doodle 15. The zonation of marine algae on the rocky shore.

We will see a vast difference in the wave-exposed rocky shore, where the inhabitants have to withstand or hide from huge pounding waves. The animals living here often grow to a smaller size and are more streamlined than if they were living in a low-energy environment. Those creatures that cannot move often have stronger attachments, and those that can move often take shelter in crevices. We find a vast reduction in species diversity; however, those species that do live here are more significant in numbers.

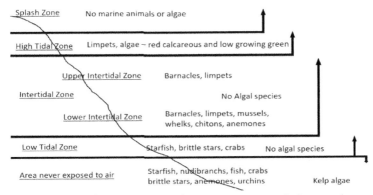

Doodle 16. The zonation of a wave-exposed shore where we can see the significant decrease in animal species diversity and also the near absence of all algal species.

As you can imagine, the food web of the rocky shore is vast and complex. Not only that, but it will differ between different localised areas.

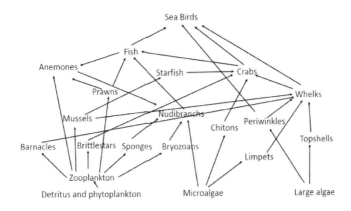

Doodle 17.

However, we can produce a generalised food web for this area, as shown in doodle 17.

Within the rocky shore and subject to localised variation, we find five distinct habitats that various animal species have successfully colonised. The first and most common is the rock pool, an area often associated with fun, falling in and nets. These are little oases of life when the tide has receded. Their size ranges from extremely small to very deep, vast expanses of submerged rock. The larger the pool, the larger the diversity of life. In large rock pools, we will find animal species that are restricted to that individual pool as they are unable to tolerate aerial exposure; these are typically mobile predators. The volume of water these large pools contain is able to keep the physical conditions stable and animals commonly only found below the low tide mark can be located here. Great fun indeed!

However, as we move to the smaller rock pools, we find more extreme conditions; the water will cool or heat quickly, but to a lesser degree than if exposed to air. The biggest problem for animals who take refuge here is the oxygen content of the water. Now, if you are adapted to being exposed when the tide goes out, that's no problem, but if you are a more mobile species, it's dangerous.

In the summer months, when we have relatively high temperatures, we find that oxygen is quickly removed from the water, and soon animals such as crabs are in danger of succumbing to oxygen starvation. Here we see a fascinating behavioural adaptation, one that's worth looking out for on a hot summer afternoon. The crab will move up to the water surface and start beating it with its mouthparts. The water bubbles and the surface moves as the bubbles break. This allows oxygen to enter the water and the crab is saved. The only risk is that now the crab has exposed himself to predators so has to be on constant lookout.

Another thing to consider when we are looking at the small rock pools is the introduction and loss of water. On a hot, sunny day we will find that there is a high degree of evaporation from the water surface. Thinking about it, only fresh water evaporates, leaving all of the salt behind. So what we have here is increased salt content in the water, which, in extreme conditions and small rock pools, could totally wipe out life there.

The second aspect here would be if it were raining; not just raining - I mean throwing it down. The substantial introduction of fresh water, which may also be slightly acidic, would dilute the pool's salt content. Although not as deadly as the latter, if we also had runoff of fresh water from rock faces into the pool, it could cause extreme problems for a short period of time.

Another habitat that we find is one associated with two significant constituents: a surface to live on and one to hide under - the algae. The algal fronds, which people often think are the leaves, provide a large surface for small filter-feeding organisms to colonise. If you actually take the time in the lower intertidal zone, you will be quite amazed at what you find growing and living on the algal fronds. Take a bucket of water and a magnifying glass and submerge the fronds; try it, and you will be surprised at what you see.

Also, when the tide recedes, and the fronds are lying down, they provide a unique habitat for larger animals taking shelter there. Anything hiding underneath the fronds is protected from the sun and rain, and lives in a damp area. Many animal species can be found, including crabs, starfish and even fish such as blennies. The holdfasts of the algae create refuges for creatures to hide between, where they can avoid predators when the tide is covering them.

Crevices are hugely influential in the more exposed areas as they provide shelter from the pounding waves and water movement. Here you will find aggregations of many mobile species such as snails, whelks and crabs, all taking shelter.

Boulders also provide a unique habitat for many animal species that are not found anywhere else on the rocky intertidal shore. They provide shelter due to their sheer size, but the community existing on the boulder is representative of that of the area - nothing unique there. However, when we look at what's known as the under-boulder community, we find something completely different. This area is protected from extreme wave movement and desiccation, and here we see a community of shade-loving animals such as sponges, bryozoans, and sea squirts. These often exist under small rocks that are lifted when rock pooling. If left exposed, they would die, which is why it is so vital to replace any stones you move to their original position.

One final area that is of enormous importance is known as the tidal rapid. Many of these are now marine reserves because of their uniqueness and endemic species (species that only exist in this environment). An area between two landmasses, the tidal rapid is protected from wave action, but because the space between the landmasses reduces in size, this allows very clean, fast-flowing water to rush over a relatively small area. Submerged rocks and fast-flowing water make this an extremely dangerous environment, where one should never swim or even navigate a boat.

Here the unique physical environment creates conditions for individual species of algae and animals. The majority of the algae are large kelps. Most of the animals are small filter feeders. The large kelps have extremely strongly holdfasts keeping them in place, and the filter feeders are small and streamlined to reduce the effect of water movement over them. Each incoming tide race brings in new and abundant food supply for these animals.

So, we now have a rocky shore that has a community structure in line with our zonation, and different habitats are contained within the seashore. We would be forgiven for thinking that this was a stable community of species, but this is nothing further from the truth. Let's think of our rocky shore as a beautiful piece of real estate; what's the one thing that's in high demand? Space. That is, space for the larva to settle in and space where growth can occur. There is enormous competition between individuals and species for this precious commodity.

So what creates space in this environment? We can answer that in one word: disturbance. A disturbance will be anything that knocks off the existing inhabitants and leaves bare rock. Predators most commonly create disturbances; however, also think of a large log scraping down a rock face. Unfortunately in our era of pollution, large plastic containers do the same, and on a significant scale, so do pollution events such as oil spills.

Furthermore, natural events can occur when the mussel is the dominant life form on the rocky shore. Here, a layer of mussels will settle and grow, and then on top of them more larvae will settle and grow. The colony gets bigger and bigger until a lower layer (on which everything is built) eventually dies. We then have a bit of a storm, and the dead lower level, which is not attached to the rock anymore, breaks off, taking all the upper layers with it and leaving bare rock behind. All these actions create an exposed rock face to colonise.

Now, in many habitats we have succession, which is the order in which areas are colonised by species. There is a predictable procession of species in the community. It starts with the pioneer species which alters the space, allowing the secondary species to colonise and out-compete the pioneer species. This goes on and on and on until we reach a stable climax community, which remains in place for decades, if not hundreds of years.

The rocky shore is different, which is why we have a great diversity of life at any one point in time. We have our bare rock, and who's going to move in? This all depends on the season and the larva species that are present in the water at that time. If one piece of rock was bared on up to 10 different occasions during the year, it could be colonised by over 20 different species, depending on which larvae were present in the water at that time. Another factor here is the size of the disturbance; the more significant the disruption, the more chance of a higher species diversity settling at one time.

So what we have is an area of land between the high and low tide zones that has had sections of disturbance of different sizes, at different times, all experiencing different stages of settlement, growth and succession. It gets incredibly complicated, and it never stops moving. That's the rocky shore for you.

Random 3.

Samantha the Serpent Star

'Some people call me sexy because of my vivid red colouration, but many of my relatives exhibit all range of colours and can be found from the shallows, to the deep, dark, cold abyssal depths. I have never seen many of my cousins but often throw food to them, off the continental shelf - well you have to do your part, don't you!

To be honest, I am quite a thug. At feeding times, I have no manners and will try and outrace anything to get to that big piece of chunky flesh. Oh yes, and I am quite the sprinter, you know. I will do anything to win, and once that meat is within my arms, I will rip it to pieces. I will then stuff in so much that you will actually see my back bulge upwards with the amount of flesh forced inside me. Fine dining - who needs that?

Yet at the same time, I can be so delicate that I can handle a single grain of sand with my feet, pass it along and play with it. I also care for my babies so much; I love my little darlings, and I won't have them playing in the plankton waiting to be eaten. I nurture them in a little sac inside my body, until they crawl out and are independent, bypassing that dangerous planktonic phase. We serpent stars are so successful in rearing our children that we can crowd the sand with as many as 1,000 of us, living in just one square metre of sand or mud.

I have a water vascular system, which is like your blood flowing through your veins. This water passes down my arms and, using a series of valves, I can extend any one of my hundreds of tube feet, each independent of each other. It's like turning on a tap when I need them and turning it off when they rest.

My tube feet are great at finding small bits of food; I can throw an arm between two rocks, extend my feet into the crevice and remove any organic material lying there - great for cleaning out dead bodies. However, be warned: we like meaty foods and will not tolerate eating bits all the time. Hey, would you be happy with scraps? Now, what's meaty that swims? Yum; yes - fish - and I will take them if I need to. Sleeping fish just don't wake up!

I was passing the house of a good mate of mine the other day, and she was not in good shape. Even with our calcium carbonate armor, we can get into a bit of trouble out here; it's eat or be eaten. (Personally, I like the 'eat' bit best.) She had been in a fight with a fish; cheeky sod had taken a bite right out of her - three arms bitten off down to the stump. However, all is not lost, as she was already starting to grow them back. We got our revenge; I collected that fish's eggs, and we both had a good meal. Ha! The energy contained in the eggs is transferred to her newly growing arms; fair swap, I say.

Marine Ecology for the Non-Ecologist

At the end of my five arms are my sensory tube feet, which I use to get around and feel the water. They tell me to retreat and run, and they also tell me when food is available. Right now, they are screaming: 'amino acids in the water just north!' Some poor fish is wounded - been in a fight, I think.

So it's goodbye from me because if I don't get a move on, that crab will get there first, and there is nothing like fresh fish!'

4. THE MUD AND THE SAND

Oh, the sandy beach! The crowded sandy beach; sunny days, ice creams, sandcastles, sunbathing and burying your dad!

Walking over the sand, we would find countless empty shells. Did we ever question where their owners were and how they lived? Spirals of sand indicated the presence of worms (again just worms!) On the infrequent occasions when we found hundreds, if not thousands, of stranded dead jellyfish, did we ever consider what would happen to their bodies?

To understand exactly what lives within the sediments and how they go about their lives submerged in the sand, we will first not look at the animals, but at what they're living in. Let's look at the sand itself.

One of the main things that will strike you when looking at a beach is its vastness. Most beaches are enormous, with the vast expanse of sand stretching hundreds and hundreds of metres from the land to the sea. As a rule of thumb: the longer the tidal range, the shallower the beach. Also, what is critical is that the more extended the tidal range, the quicker the tidal race when it turns and starts to move up the beach; on some occasions, faster than you can run. Unfortunately, this fact alone has led to the deaths of many people who were caught out by the incoming tide, not understanding the situation they were in. This was highlighted one year in Morecambe Bay in the UK, a dangerous place indeed, when 21 cockle fishermen, who were not local and didn't know the area, tragically lost their lives as the tide swept up around them. It came in faster than they could run and caught them.

So we stand at the high tide mark and look down towards the sea, over the bare, lifeless expanse of sand. However, we are not looking in the correct way if we are to consider the life that is hiding there.

First, we must consider the grain size. Yes, the actual size of the grain is hugely important. It will govern how water moves through the sand; it will dictate how easily the animals can navigate and push through the sediments; and it will also determine how much food is contained within the sand itself. The latter statement is incorrect as the sand itself will not hold the food, but we must look at the spaces between the sand grains instead. These extremely important spaces are affected by the size of the individual grains - get the picture?

Now there is a whole science that has studied the size of the grain and the spaces between them. We shall leave this extremely complicated area and simplify things by identifying what happens with only two different sizes.

Doodle 18. How grains of sand lie above each other, interspersed by spaces that contain salt water and many other things.

We will consider two types of sand. When you step on coarse sand, the water is pushed away from the site of disturbance, leaving the sand more compact and difficult to penetrate. This is the natural type of sand that we find on a beach. However, where the particles are smoother and smaller, we see the opposite happen. Here, water will move towards the disturbance, making the sediment more fluid. However, when we remove the interference, we find the sand hardens, making it difficult to extract anything. I wonder how many rubber boots have been lost in this environment?

Again, this could lead to hazardous conditions because it is not possible to identify the type of sand simply by looking at it. Back to Morecambe Bay, where there are so many areas of quicksand hidden within the smooth walking sand that it is easy to get into trouble very quickly. When we consider how fast the tide can move in, we have potentially life-threatening situations. The danger is such that you can only cross this expanse of sand with a local guide.

There is another hazardous event that can take place on the sandy shore and in localised areas of fast currents. In fact, people have been swept out to sea and lost in these conditions. What we're talking about is water movement underneath the sediments when the tide is racing in. This water movement, combined with the pressure of your feet on top of the sand bed, causes the sand to collapse under your weight. This creates a complete dislodgment, a loss of balance, and before you know it, you're taking a swim. It is more exaggerated on the steeper sandy shore; the steeper the beach, the deeper the water, so within a couple of metres you have slipped into chest-deep water and are in danger of being swept away. As always, when on the shore, be careful.

I digress here to mention just one particular sand grain; not only is the size of this grain essential, but so is the material of which it is made. There is a very rare beach indeed, where we find the two physical properties in perfect balance. These are called the whistling sands, because when you walk over them and compact the grains, they whistle a tune. It is a bizarre phenomenon and something that everybody should try if given the chance. But let's get back to ecology.

So, we have water in between the grains, and it's not just sea water but a hugely complex habitat in itself. Over the sand grains we have a thriving population of bacteria - our decomposers, while within the water we have a whole community of microscopic animals. To add to this, there are larger pieces of broken-down bodies on which all of these are feeding, commonly called detritus.

We now have the sand, of various grain sizes, and the makeup of the water between them, as well as what is living on and between the sand grains. In other words, we have our house for the animals that live there. For humans a house gives us protection from the elements while keeping us safe and warm, and this is precisely what the sand provides for our animals.

When the tide is out, the sand offers protection in many ways. If it's raining, we find that the fresh water simply flows over the surface and will not penetrate the sand, so the animals are protected from changes in salinity. If we have a scorching day, the sand reflects the sun's rays and reduces any form of heat shock that the animals may suffer. It also confers protection by hiding them; however, many predators such as birds have evolved different beak types that allow them to penetrate the sand and find their prey.

The sand will cause a particular problem for the animals living there, because the deeper into it we go, the less oxygen we will find. In very coarse sand, oxygen depletion occurs at a greater depth, but in the case of a tiny particle such as mud, oxygen is lost extremely quickly and may be totally absent within a few millimetres of the surface. However, with each incoming tide, the oxygen is replenished.

When looking at a beach, we can consider two communities: the surface community, and the bottom community. The surface community can be split into two: the mobile and the sessile, with the two major sessile communities being governed by the makeup of the beach itself.

If we find a beach that has a few sizeable stones or small boulders, we will discover algae attaching to them, as well as barnacles and sessile shellfish such as mussels. These live in small, isolated static communities. However, if the algae grows too big and we have a storm, we find that it can either be ripped off the stone or the stone can be lifted and moved up and down the beach with the waves. If the stone settles too high up the beach, the algae's successful growth is its eventual demise. If it settles too far down the beach, the algae may not receive enough sunlight and again succumbs.

The other more massive community is better described as a habitat that is rarely exposed to the air. It exists in shallow water and is known as the sea-grass meadow. Here we have a unique plant, for it is the only true marine plant. The difference between a plant and an algae is that a plant has a root system serving not only as an anchor but also to acquire nutrients. Also, a plant produces flowers. All species of algae are totally absent of roots and flowers. So here we have the only true marine flowering plants, of which there are 49 species.

The roots of the sea-grass penetrate and stabilize the sediments while their grassy blades slow the water flow through them. This allows more detritus, which will then rot down and feed the plant, to fall out of suspension. As with all things marine, seagrass will provide that hugely important piece of real estate called a surface area. This allows a diverse community of small filter feeders to settle on the blades and fulfil their life history. You will also find small species of urchins, brittle stars and small fish living within the leaves.

The food web of the sea-grass community is quite complicated, as you would expect. However, when we look at the base of the food web we find there are three different food sources. Here we have the primary producers, which in this habitat are the phytoplankton and the seagrass with its roots. Furthermore, we have algae and a considerable amount of detritus adding to the basis of the food web.

Phytoplankton, which the sponges and all the filter feeders will consume, forms the basic food source. The algae and detritus feed our small grazers and decomposers. Although sea-grass itself is too tough to provide much nutrition, its leaves are important as they break off to form the detritus bed. Additionally, its roots provide valuable sustenance for many animals, including a favorite among many conservationists: the manatee.

However, as these habitats are mostly inaccessible from the coastline, we will leave the sea-grass and their amazing communities here.

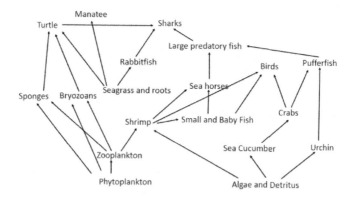

Doodle 19. The seagrass food web with its three-tier base.

The main predators that swim up and down with the tide are the mobile community of the sandy beach. They are fish, with small flatfish and the juveniles of all the species at the smaller end of the size range. These opportunistic fish will feed on any meat, whether it is dead or alive. We then move into the arena of more substantially-sized beasts such as the large rays, large fish, small sharks and, in tropical areas, huge sharks indeed. These glide in, looking for things to scavenge, and they actively hunt the smaller fish, as well as anything that presents itself on the water surface. When the tide goes out, these predators retreat. All these species can be found in only a few feet of water and very close to the shoreline.

Now we shall move to the bottom-dwelling communities of animals that live within the sand. What we find here is that many animal species exist in localised populations, resulting in some areas of the beach showing a great abundance of life, while other regions are much more sparsely-populated.

The four major groups of animals inhabiting this area are the worms, shellfish, crustaceans, and echinoderms (urchins and starfish). We will see a zonation of species associated with the aerial exposure of the beach. The high tide zone is characterized by the layer of flotsam and jetsam called the strandline; this is basically made up of anything that floats, such as driftwood or plastic, but is mainly broken-off sea algae. While the top of the intertidal zone is mostly devoid of large life forms, in the lower intertidal zone we will find worms, shellfish and crustaceans, with the echinoderms being limited to the low tide zone.

When we look at the strandline, we will see that the broken algae's surface is dry and crisp, looking totally devoid of life. However, if you lift a piece of this, you will discover a damp environment on which many types of amphipod, like woodlice, are living and feeding, breaking down the organic matter to be recycled. In hot summer months, you will also find the area an oasis of life for flies; in regions of giant deposits, you won't be able to spend much time due to the flies buzzing around your head. Happy days!

Now, we have a few types of worm living within the sediment, the most famous being the lugworm. Although unseen from above, their presence can be indicated by a shallow depression 10 cm away from a mound of sand that has been pushed up to the surface. The shallow depression is the head shaft where the worm sucks in water. The replenishment of water and its organic content allows a healthy bacterial population to thrive on the sand; this is the worm's primary food source. It ingests the sand, digests the bacteria and then deposits the clean sand on the tail mound. You will always find populations of worms living in an isolated area.

When walking over the sand, you will also notice localised regions where tubes of sand grains seem to be sticking out of the sediment. These are parchment worms, and such is the aggregation of these tubes that they can be considered to be mini reefs. The worm uses a special glue, with a unique chemical makeup, to stick the sand particles together. Grown worm larvae will only settle out of suspension when they can sense the glue's chemical composition and thus settle in an adult community of their own species. This increases the patchiness of the species over the beach.

There is a highly mobile and predatory worm existing in this area known as the ragworm. Think of a centipede, increase its size, allow it to live in a marine environment, and we have a ragworm. When the tide floods, it will move out of its burrow and actively hunt anything smaller than itself. (Unless of course, it meets a fish and becomes its prey.) The largest of these is the king rag, attaining a length of over 30cm and possessing a formidable pair of pincers that are capable of inflicting great pain on anyone who is unaware.

The sediment-living molluscs are mainly bivalves (two shells stuck together) and again exhibit patchiness over the beach. Their zonation can also be vertical, as some species bury themselves deeper than others. The two most common are the cockle and the razor shell. All of the species move within the sediment by squeezing out a sizeable muscular foot and pressing it against the sand. They are all filter feeders, and they stick out a hollowed tube called a siphon, taking in water containing plankton and dissolved organics.

The cockles, which live higher in the sediment, have much more robust and rounded shells that offer some protection; but many birds can dig them up and make an easy meal of them. The deeper razor shells have a more streamlined shape as the sediments are more compact, and their shells are thinner as fewer predators are able to reach them.

There is always a large population of mobile crabs that are able to bury themselves quickly within the sediment to get out of trouble. The predatory crabs will have a sizeable crushing claw to break open shells and a thinner cutting claw to extract the flesh. There are many species of crabs that you will not see as they remain buried for most of their lives, only to stick out feathery antennae when the tide is in, to filter food from the water column.

Then as we go to the lower intertidal zone, we find our brittle stars, starfish, sea cumbers, and urchins. Again, unless you want to dig them up, you won't be seeing them. However, the urchins and sea cucumbers are of enormous importance when it comes to the ecology of the area.

The sand bed will soon go anoxic if it's not regularly cleaned, and here we enter the world of bioturbation.

This is the turning over and cleaning of organic matter from the sand. Here, the urchins and sea cucumbers really play their part; they are like substantial underground bulldozers, moving through the sand, consuming all the organics and then depositing clean sand behind them. It has been found that individuals of certain urchin and sea cucumber species will process over 90 kg of sand per square metre of seabed every year. When we consider that many hundreds of individuals will inhabit a single beach, this is cleaning up on an industrial scale.

The primary physical factor that causes a biological disturbance on the sandy beach is stormy weather. Now, if we have a particularly vicious storm combined with a spring tide, we will have relatively huge waves breaking over the beach. The height of these waves will be higher than the water depth, causing the water to penetrate the sandy sediments and dislodge any animals buried there.

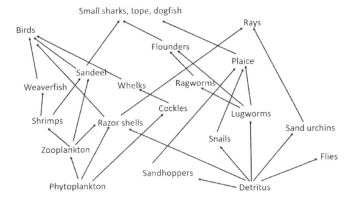

Doodle 20. A simplified food web for a temperate sandy beach.

Just behind these breakers are the large predatory fish, waiting for the dislodged animals to be presented to them. This action can cause the complete removal of a localised aggregation of a single species; the patchiness has gone. Happy times for the fish! And even happier times for any fishermen willing to brave the weather, casting a line into the surf and hoping to catch a big one. Once the storm has passed, the sediment returns to its normal state, and the larvae settle to form a new community. Patchiness returns, but it is not necessarily the same species that will inhabit the cleared section of sediment; the community is as fluid as the sand it inhabits.

Random 4.

The Empty Sea

Is the sea emptying of fish? People have known this for many years and ignored it.

'It will go away won't it?' they think, with their heads in the sand. No, it won't. But it is not too late.

The idea of fish stocks being in total collapse has been around for many years. Let's look back at an important

event on the social calendar in 1930's UK. Champagne-sipping socialites migrated from their ivory estates in London to make an annual pilgrimage to the Yorkshire fishing town of Whitby. There, the little darlings went out in small rowing boats dragged by larger boats; soon they were afloat, alone, fishing rods in hand.
'Tally-ho old chap!' they shouted, as the mighty bluefin tuna dragged the small boats out to sea. Magnificent beasts they were, growing to over 500kg - what a sight it must have been. But where are the socialites now? They haven't been to Whitby since the early 1950's.
'Sorry old chap, no tuna left anymore. All gone.' But why?

Simple: all the North Sea herring stocks, the tuna's food, have been scooped up by nets. Food for tuna gone = tuna gone.

Fish stock depletion has been going on for longer than you think. Too many years ago now, I was a 'mature' student having a pint with my fisheries professor and was enlightened on the political merry-go-round of the day. He told me the government paid scientists to calculate the fish stocks in an area – let's say cod had a stock of 100 tons. They would then estimate how much they could take to allow a healthy breeding stock - the fishing quota. Let's say it was 25 tons of cod. The politicians of the time then took the scientific information they had paid for and did what they did best - ignored it. The UK politicians needed a deal with

France so they gave France 20 tons of cod quota. The UK fishermen were unhappy with this, so they were appeased by being given a quota of 20 tons as well. Now, call me stupid, but the math just doesn't add up. Multiply this by many years and many species, and we have an alarming situation.

These are just two simple examples. Unforgivingly, there are too many instances of mismanagement currently affecting the fish stocks. Political decisions, illegal activities, loopholes and basic exploitation are working synergistically to destroy our fish.

Yet it is not too late. **FISH STOCKS DO HEAL.** This is a fact. A good deal of science has shown that no-take zones are the best way to repair the damage; not only do the stocks recover and begin to thrive, but they also do so quite quickly. They heal, and we can reverse this graph. In every country in the world, fantastic organisations exist, staffed by unsung heroes, working on this goal. Large or small scale - it doesn't matter. Every no-take zone helps. And this is only one instance of many potentially helpful solutions. You could write a book on it. The simple fact is that fish stocks are in serious trouble. It can be reversed, but only by people working together.

Close your eyes and think of 100 gold coins piled up - it's easy. Next, think of and visualise 1,000 - it's harder. Now, imagine 1 million coins. You cannot do it - it is impossible to envisage that number. 100 million sharks, not 1 million, but 100 million sharks are killed each year for their fins! It makes you think, doesn't it?

5. WHERE RIVERS MEET THE SEA

Rivers meet the sea where fresh water, often flowing downriver for hundreds of kilometres, finally leaves the terrestrial environment and mixes with the salty waters of the ocean. A salt concentration gradient exists within the estuary's waters, with saltiness gradually decreasing as we move up the inlet (and often into the river itself), until it finally disappears. This area is commonly known as the brackish-water environment.

All estuaries, no matter where you are on earth, are very new in geological time. The reason for this is that they were all formed as a result of the ice melt, with the last Ice Age ending just over 10,000 years ago. They are very fluid in nature because of the constant deposition of sediment and its colonisation and stabilization by different plants inhabiting the shores.

Estuaries vary in size from the tiniest tidal creek, meandering its way to the shore, to the lower reaches of a massive river such as the Amazon, whose influence can be observed many miles out to sea.

Four primary types of estuary have been characterized based on their surrounding landscapes and the processes that formed them.

The most common is the coastal plain estuary, and this is what most people will think of when asked what an estuary is. This type of estuary was formed when rising seawater flooded the low-lying coastal river valleys, pushing the rivers inland. If you contemplate that sea level increased by around 125m during the ice age melt, this would clearly push a river mouth far inland.

The bar-built estuary is typically very shallow and protected by a wall of shingle or sand that is deposited parallel to the shoreline. This reduces the sea's influence on the estuary and allows freshwater content to push further out into the estuary's mouth. It also protects the life within the estuary from storm surges and violent waves.

The fjords are a result of enormous glaciers carving out the valleys during the Ice Age. These are commonly very deep and narrow, often over 100 m deep just 20 m from the shore and frequently over 500 m deep. They all have at their mouths a shallow submerged reef, which is composed of the glacial deposits (large boulders) which were left behind when the ice melted. This bar at the mouth significantly reduces water circulation in the deeper areas, resulting in fast-moving currents in the surface waters and slow-moving currents in deeper water.

Tectonic estuaries occur in small numbers and have been created by earthquakes along a fault line. Here, large sections of land have subsided into the earth's crust, allowing the ocean to flood the area. Where we have a river located in this landscape, we have a newly-formed tectonic estuary.

The three major physical factors that apply pressure to any of the animals living here are water circulation, salinity and sedimentation.

Water circulation and salinity can be looked at together as they are closely related in this environment. Fresh water weighs less than salt water, with the salt water being denser. This means that when we consider the estuarine environment, we find two layers of water. The freshwater input from the river flows over the surface towards the sea whilst the denser salt water penetrates the whole estuary in the deeper regions. The way these two water bodies mix determines the extent of the brackish-water environment. If we start with unadulterated sea water and move up the estuary to pure fresh water in the river, the area in between is the brackish environment, where salt content decreases towards the river.

Get two half-full, clear glasses of water and add a few teaspoons of salt to one, mixing until fully dissolved. Then gently pour the salt water into the fresh water while observing the water itself. Here you will see how the two different types of water mix; you will notice the salt water going towards the bottom, and you will see the boundary between the two different waters as they slowly mix together. It's simple to do, but fun.

The same happens in the estuary: the salt water will eventually mix with the fresh water, but how it blends and how long it takes to mix depends on a number of highly varied factors. These include tidal velocity, the volumes of fresh water, the amount of sediment, the local topography or shape of the estuary, plus other local factors. To add to this, we have regional variations in weather. Has it been dry - reducing the fresh water and sediments? Or have we experienced a high degree of rain, causing vast volumes of fresh water with significant deposits to be discharged? All act together to increase or decrease the amount of water-mixing in the estuary.

There are three different ways that water mixes within an estuary. The first is the salt wedge estuary, where we have minimal mixing between the two water bodies. This occurs where there is a significant freshwater input in relation to the tidal flow. The salt water from the tide penetrates high up into the estuary in the lower depths, while the fresh water from the river flows over this and out towards the sea. There is minimal vertical mixing of the water, and we can see a very distinct line between the two; this marked difference in salinity is called a halocline.

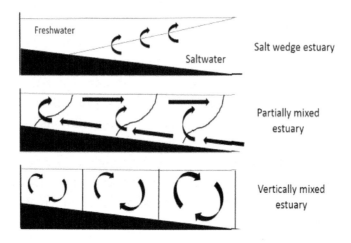

Doodle 21. The different types of water-mixing within the estuarine environment.

As we move into a different estuary, with a small river discharging less fresh water, the tidal flow increases relative to the freshwater input. Here we find a two-way vertical mixing of the water bodies, which destroys the wedge. A higher degree of salt water is moved towards the surface and mixes with the fresh water. However, we still have a distinct bottom flow of sea water and a top flow of fresh water, but the surface waters have a higher salt content than those of the salt wedge estuary.

Lastly, we see the vertically mixed estuary, where the freshwater imports are low, and we have a strong tidal influence. Here the tidal rush is so intense it vertically mixes any fresh water that enters the estuary. We see the same salinity at the bottom of the estuary as is recorded at the surface. As we move from the ocean towards the river mouth, lower salt content is evident, so salinity decreases in the horizontal plane when moving away from the sea while remaining the same vertically throughout the water column.

If the majority of people were asked to describe the surface of an estuary, they would say it was sand or mud, and correct they would be. Sedimentation is an important factor within the estuarine environment, with estuaries generally being covered in fine thick mud. To this extent, most estuaries exhibit what many people would call 'dirty water'. That is, the water column is full of fine sediments with very little light penetration. However, some estuaries, such as fjords, can be clear after a dry spell of weather. Yet, following a prolonged period of rain, peat is washed off the land into the water, which can completely discolour the whole fjord within 24 hours. Nevertheless, some remain crystal clear all year round!

Sedimentation processes are incredibly complicated; however, all follow roughly the same behaviour. We have a situation where extremely fine particles are transported by the river to the sea. Now, when these particles enter different water chemistry, something changes on the particle surface; on the molecular level, their electrical charge alters. So instead of repelling each other, two small particles will attract each other and stick together. As more and more particles stick together, they become more significant and substantial until they are so massive that they start to fall out of the water column and settle into the sediment in a process called flocculation.

Now, depending on where we are in the world, we find that every incoming tide will bring to the estuary organic matter in various degrees of decay - mostly broken-off sea algae and deceased animals. As the tide recedes, these will be left upon the estuary floor to be mixed with the fine sediment that the fresh water deposits every second of every day.

So when we consider an estuarine food web, we will see that the water itself is highly turbid with hardly any light penetration and is high in decaying matter. Instead of the food web being based upon primary producers, it is based upon the availability of detritus and its associated decomposers.

Other notable factors that influence the estuarine environment include temperature. Here, animals will experience a more rapid and higher degree of temperature fluctuation than in the sea. This is due to the relatively small volume to surface area ratio of the water body within the estuary but, more importantly, to the higher degree of temperature variation in the water discharged by the river.

Because we have an environment that is very high in decaying particulate matter, we have vast populations of bacteria acting on this food source. Now this population of bacteria is extremely efficient at stripping oxygen from the water and the sediments. In some places, we could have total oxygen depletion within a millimetre of the mud surface.

So we have a dark, turbid water column that's full of sediment and detritus, the salinity decreases with distance from the sea, while the vertical mixing of the water depends on where the estuary is in the world. We also have a low oxygen content with high-temperature fluctuations and a thick constant deposit of mud 24 hours a day. Not much for the animals to cope with, really! Not only do they cope, but those that can live there thrive, as estuaries are highly productive areas indeed.

Let's look at the life forms existing in this environment and at the type of zonation we find here. Zonation is not restricted to the intertidal zone as on the rocky shore, but here we find it is governed by the salt content of the water. We see three different groups of animals: the saltwater species, the brackish-water species and entirely freshwater species.

We also have to consider one hugely important factor - time. It is accepted that it takes between 2 and 20 million years for a species to evolve, and all the estuaries in the world today are only 10,000 years old, so we have a unique situation. Here we have a habitat that has not been around long enough to be adapted to and inhabited by a wide range of species. So what we see is low species diversity; but those species that do inhabit this area, do so in extremely high numbers.

The zonation starts at sea with the salt content of $35^o/_{oo}$. Here we find all the genuinely marine species, which are called stenohaline. That is, they can tolerate only a very narrow range of salinity change, and all species are restricted to the lower reaches of the estuary. Then we move into the brackish water, which is characterized by a salinity range of between $34^o/_{oo}$ to $5^o/_{oo}$. Here we observe the animals that can tolerate a high degree of salinity and are termed euryhaline. These animals inhabit the greater part of the estuary. With salinities lower than $5^o/_{oo}$ in the upper regions of the estuary, this is where we find the truly freshwater species.

So if we start at the mouth of the estuary, we find the full species richness we would expect on the sandy shore. However, as we move into the estuary, many species are quickly lost. Starfish, urchins, sea squirts, algae and many more do not penetrate into the estuarine waters.

As we move into the brackish-water environment, we suddenly find our species richness dissolving into the fresh water. We see mostly benthic invertebrates like worms and shellfish, with crustaceans limited to certain species of crabs and isopods. Algae are generally deficient in this area due to the low volume of hard surface to attach to and also the low light penetration when covered in water.

As we move up into the river, we see a total exclusion of brackish-water animals, which are replaced instead by entirely freshwater species. However, the salt water influence can penetrate surprisingly far upriver, increasing the presence of brackish-water animals in the river itself. It might be uncertain whether this part of the river is indeed a river. Yet, when we look at the salt content of the water, we see that the brackish-water environment can have an influence over a kilometre from the river mouth.

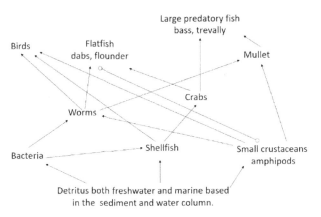

Doodle 22. A typical food web in the estuarine system. Note the distinct lack of species compared to other habitats.

When we observe the high densities of the populations associated with the estuary, we are actually looking at a vast amount of food for predators. Estuaries are, therefore, significant areas for migratory birds to stop off and recharge their batteries before continuing on to their destinations. We also see that the flooding tide brings in new predators in the form of fish.

The prey species achieve a spatial refuge by inhabiting the higher intertidal zone and upper reaches of the estuary, where predatory fish cannot seek them out. However, at low tide, these intertidal prey species are consumed by the bird population, so their spatial refuge is lost. Often the bird numbers are so high that they can decimate an area, leaving not one single prey item alive.

Again, this leaves space for new larvae to settle and start to grow, so very soon we will have a new population of animals. It might be a different species of worm or shellfish, growing and thriving in that area, although their time is limited until the next bird migration.

Random 5

SAR 11: the most abundant life form in the ocean

What's the most important marine life form?

Marine Ecology for the Non-Ecologist

Well, that's a question, isn't it - what do you think?
Whales and dolphins will spring to mind for so many
people, but I did not ask what the most intelligent marine
life form is. So what is it?

Is it coral? For without it, reefs would not exist, and our
landmass would look totally different. Or is it algae: the
basis of 99% of life in the oceans? Well it could be, but
another organism is also in contention. One that possibly
has more effect on carbon dioxide levels in the sea
than anything else – so is of huge importance.

These beasts collectively have a direct impact on the pH
levels of the world's oceans; they are
incredibly influential on carbon cycling in the sea and
are significant players in the production of
oceanic methane (a greenhouse gas 20 times more
effective than carbon dioxide). Our beast has no name -
it's only known as SAR11, the most abundant marine
organism known.

Its combined weight will exceed all of the fish in the
ocean - think of that! Yet, a single millilitre could
contain 500,000 individuals. Here we have the smallest
marine bacteria discovered, but the most abundant
marine life form known.

So we have a saturated sea of simple floating bacteria,
which feed by taking in carbon compounds and

converting them to other organic compounds, producing carbon dioxide as waste. We can see how extremely important these bacteria are for maintaining the carbon cycle.

However, they don't just need carbon to function; they also need phosphorus, and when this is in low concentration they turn to a compound called methylphosphonic acid. The breakdown of this compound produces methane gas, which is now in a saturated state in many areas of the world's surface waters.

So our beast is the most abundant life form in the oceans, (and we can't see it!). It is the essential player in carbon cycling, while its importance in methane production is only just being uncovered. Yet it is being destroyed at an incredible rate. No, no, no - for once, we humans might not be to blame. Instead the culprit is the bacteria's age-old nemesis: a killer virus. It latches onto the bacterial cell, which it infects and kills. The dead cell then splits and sinks into the abyssal depths, taking all that carbon with it – more carbon cycling to understand! One great thing here is that our bacteria is always one step ahead; every time the virus mutates to become more active, the bacteria develops another defence to trump it.

We find that the most biologically relevant beast in the sea could be the simplest bacterial cell known. It exists in mind-boggling numbers that outweigh all the fish and is in a constant survival battle in the biological war for supremacy over the microbiological world.

Next time you accidentally take a sip or mouthful of seawater, think of how many SAR11 bacteria you have just taken out of the equation!

6. THE MARSHES AND THE MANGROVES

A unique habitat exists on coasts around the globe in an exceptional area between the high water spring tide mark and the high water neap mark, (rather than the true intertidal zone). However, for it to exist we also need something quite unique to happen: the mixing of entirely terrestrial soil and sediments of a marine origin. With this mixing, as well as daily flooding of either saline or brackish water (depending on location), we find the creation of a unique sediment. Only in places where the deposit has just the right grain size and is bound together by high organic content, will our habitat develop over time.

In the temperate to sub-polar regions we find salt marshes, while moving towards more tropical zones these are replaced by mangrove communities. Specialised terrestrial plants form the basis of these communities, and they live in an area that no other plant can colonise. These plants are termed halophytic; in other words, they are salt-loving.

A sure way to destroy an area of land is to flood it with salt water. This has been apparent in many tropical regions where farmland was inundated with sea water in order to cultivate prawns. This aquaculture technique proved very successful; however, when prawn prices fell the farmers decided to reclaim their land and grow crops again. Much to their dismay, when they drained the land, their crops failed due to the high salt content of the soil. Sadly, this practice has ruined hundreds of thousands of hectares of what was previously very fertile land. To the farmers' disgust, the companies that had persuaded them to flood and had bought their prawns just moved on to other areas. This left the farmers and their communities in substantial financial trouble and unable to grow food!

The Salt Marsh

The salt marsh is usually an area of flat land which drains slowly. Stable sediments are essential as they encourage the development of root systems that will stabilise them further. For this to occur, we need a group of exceptional organisms: bacteria, diatoms (a type of single-celled phytoplankton) and filamentous algae. These grow over the surface of the sediment, binding it together with secretions and also slowing down water movement over that surface. This allows more detritus to settle, which adds to the mix to form a stabilising mat. It has been shown in experiments that if you scrape off this surface mat, very soon the whole sediment structure starts to slip and slide away, taking everything with it.

We also have a situation where the base is studded with relatively large broken shells and stones. Here again is a surface that will allow small macroalgae to colonise, promoting the deposition of detritus and a build-up of stable sediment.

Once this mat has formed, the sediments become stable enough to allow the germinating seeds of any plant able to tolerate this area to take hold. We find that the first plants to colonise are annuals; this means their lives are but one year. However, when the plants die, their leaves will add to the sediment, binding it further and supplying nutrients for the next succession of plants.

The salt marsh is ever-growing, getting bigger and bigger and bigger every year, which means that we have an estuary that is getting smaller every year as the marsh encroaches into the intertidal zone. We find here another zonation, extending from what is known as a salting cliff inland until we reach the fully terrestrial area. As we move along this zonation, we will see different species of plants or grasses, with varied diversity.

Plant diversity is low in the younger marsh, yet as we move up the marsh, where less and less tidal influence occurs, we find a greater diversity of plant life. Not only is variety greater, but plant cover is too. The higher degree of plant coverage leads to an increased amount of plant detritus falling on the sediment, which in turn creates a peatier substrate, stabilising the area. Additionally, more nutrients are provided for new plant species, stimulating a higher rate of plant growth. Not only this, but as the plants develop the soil, we tend to find it rises in height over time; this rise will accelerate the drainage of the area, further stabilising it while reducing salt encroachment.

As the soil becomes more stable, it also becomes drier. Water has to go somewhere, other than just seeping through the sediments, and as on land, where mountains and hills drain into rivers, so the higher ground of the relatively flat salt marsh drains into the creek. This creek cuts deeply into the sediments and can be an incredibly soft, muddy area, as it rarely dries out. It is a very dynamic environment, but the constant water drainage and tidal flooding causes the banks to become extremely susceptible to mudslides. Not only that, but we find that the burrowing of organisms into the sides of the creek destabilises the area further.

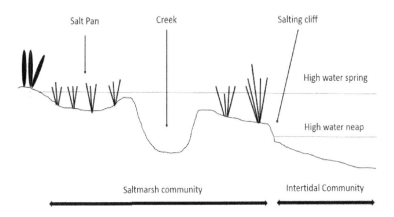

Doodle 23. A cross-section of a typical salt marsh.

There is a greater diversity of terrestrial animals in salt marsh and mangrove communities than in other marine or brackish environments. In the upper regions and in the salt pan area, we will find species of mice, rats and voles. A sizeable invertebrate collection is also present, escaping the flooding water by climbing up the blades of grass and then descending when the water ebbs. All over the salt marsh, as you would expect, a high number of birds take advantage of the bounty that can be found.

The highest diversity of life exists within the creek: the life that swims up and down with the tide, that exists in burrows and only emerges when flooded or lives entirely within the sediments.

Here we find, unsurprisingly, a dense population of shellfish. They filter-feed when covered with water and then close up house to protect themselves from any physical changes when the tide recedes. Animals that exist in burrows include species of crab, shrimp and prawn, actively feeding when the tide is in and retreating to safety as the tide drops. We also have an ingress of more mobile predators, the fish species, with each tidal flood. Depending on the size of the creek and the marine environment adjacent to the salt marsh, these can range from juvenile fish to quite large predatory animals. Some species of fish such as gobies and blennies will spend their entire lives in the creek.

In the lower regions we find mussels attached to plant stems, again further reducing water flow and encouraging detritus to settle. On the sediment we will discover mobile snails, actively feeding on the bacteria and filamentous algae that exist there. Taking advantage of any empty shells, we will find a healthy population of small hermit crabs.

The Mangroves

Sadly we live in the age where humans are destroying, at an unprecedented rate, many habitats in order to acquire whatever resources reside there. This often leaves a sacred landscape unrecognisable compared to the biological balance that previously existed there. Mangroves suffer from the same deforestation that we see in the tropical rainforest. Hardwoods are produced in mangroves, with the tallest species over 30 m high, representing a valuable resource for any unethical organisations to exploit; which they do.When the mangrove is removed, the primary organism that stabilises the sediments, provides shelter for animals as well as space to grow upon, will no longer perform that function. The tides move in (as does rain where tropical rainforests have been removed) and simply wash away the sediments. With this removal, we see the whole area changing shape, not only physically but also in its biological structure. Remove the mangrove, and we alter the biology of the area considerably and irreversibly. A new community will emerge, but it will be less diverse.

As we move into the subtropics and tropical regions, we find the salt marsh replaced by the mangrove thicket. Most people will associate mangroves with large estuaries; this is not wrong. However, mangroves exist in all areas, including tidal lagoons on coral islands; in fact they exist anywhere we have a lower energy environment and the correct sediment deposition.

As we've seen, mangroves can penetrate inland in some cases and many miles along an estuary. So we have scarce plant species that can tolerate full seawater salinity, moving along the gradient to the fresh water nearby. They can also live and grow stronger in total absence of salt. As always, when looking at the marine-terrestrial interface and its associated plant life, we find a very low diversity of species, but those that are present are found in incredibly high numbers.

An interesting fact about mangroves is that they follow the same global distribution as corals. That is, we find the highest diversity of species in the Indo-Pacific regions. As we move around the globe to the Caribbean, we see the number of species, just like corals, are vastly reduced.

The mangrove's zonation is highly dependent on its regional location. Considering that a coral island may have a tidal range of one metre, we are going to have an extremely thin area for the mangrove to colonise. Thus, no zonation occurs. Mangrove zonation is mainly found in the estuaries, where we have shallow banks. As on the sandy shore, a shallow area allows the tide to expand quite a distance. So although its range is not that high, the tide is going to cover a great deal of land with each flood and ebb. This is where we find a zonation of mangrove species. This zonation is also accompanied by different adaptations to the mangrove root systems.

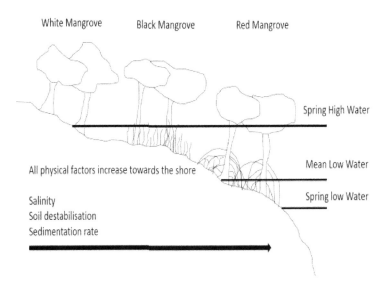

Doodle 24. A generalised zonation of the mangrove with its different root adaptations in a typical tropical estuarine environment.

The zonation of the mangrove depends on many environmental factors, all acting together.
First, we shall look at salinity. Although we know that mangroves can exist in the total absence of salt, their species show a distinct ability to tolerate different salinity levels. While some are totally marine, others can only exist in lower-salt environments. So we have a zonation of species along the salt gradient exhibited within the water and the sediment.

Climate, particularly rainfall and temperature, can also have an effect. Considerable precipitation in some regions can destabilise the sediment, as well as reducing its salt content. In areas of low rainfall, we find the deposit can contain extremely high salt concentrations, thus restricting the number of species able to colonise there. High temperatures acting on the sediment will boost evaporation rates, which will once again increase their salt content. However, more importantly, as the temperature within the sediment increases, so will bacterial activity. This may lower the sediment's oxygen concentration and in extreme conditions can lead to completely anoxic conditions. Again, the ability of the mangrove to thrive in this anoxic environment through its adaptations will produce a distinct zonation of species.

Tidal behaviour is extremely important for the propagation of the mangrove. We find that an area exposed for too long could thoroughly dry out. Any seedling here that is not adapted to dry conditions will only live for a few days after germination. This excludes and includes different species from the individual and localised areas.

We also have disturbance factors and, depending on location, some are more abrupt than others. In the Caribbean regions, we find mangroves can be susceptible to hurricanes. This can lead to complete removal of the thicket. Often, mangroves re-colonise the salty areas; however, their growth in higher regions can be stifled by faster-growing ferns that out-compete any mangrove seedlings for light. If the sediments are unstable after the removal of the mangrove, then we find the tide will erode the area and the mangrove may never return.

When considering a localised area, we can find variety within the mangrove thicket, such as salt plains or deep creeks meandering through the forest. Mangrove zonation is an extraordinarily complex and fluid topic.

The mangrove tree exhibits different adaptations that allow it to succeed; these are so extreme that it is essential to examine and understand them. If we look at the area the mangrove inhabits, we notice it is salty and lined with fine sediment. This is covered with water and exposed to high temperature and sunlight every day. The tree itself not only has to withstand the salt but must also live in anoxic, waterlogged sediments.

What most people notice are mangrove root adaptations. Here we will examine two species: the red mangrove and the black mangrove. The black mangrove lives higher up the tidal range and thus will experience anoxic sediments. Here we find root adaptations growing up from the main root system and extending into the air. The scientific terminology for this type of root is the pneumatophore.

The red mangrove thrives in the lower regions; here this plant lives in an area of destabilised sediment. Its roots extend from the trunk downwards and into the sediment, effectively propping up the tree and also helping to stabilise the sediments it is living in. Surprisingly, these are called - you've guessed it! - prop roots.

Now, both root systems are covered in small pores which exhibit a one-way system. They do not allow water in, so when covered in salt water, nothing happens. However, when exposed to air, they allow the passage of oxygen into the root. Thus, anoxic conditions in the sediment are of no consequence to the biological activity of the plant.

Salt concentration and the movement of water and salt is an extremely complicated biological topic, so to keep things in context, there are three main ways in which the mangroves are adapted to life in the salty environment.

The first is at the root level where, quite simply, there are salt gates, which close to stop any salt in the water from entering the plant. The second is at leaf level, where there are unique pores; the leaf concentrates the salt in a specific area, and it is then excreted through the pore. (It would be like crying salt water.) The third adaptation we have is the art of salt storage, where the salt is filtered from the water within the plant, passed into the leaf, and packaged into a biological parcel. Here the salt remains separated from the biological activities of the plant until the leaf itself is shed, taking all the salt with it.

Another fantastic adaptation is found in their reproduction. These plants live in soft sediment, and within the tidal influences; so shedding seeds is a precarious business, as most would drift away or be eaten. However, some species do reproduce in an opportunistic manner, allowing their dispersal over vast distances along the coastline and even to isolated islands. The best adaptation comes from species that drop their seeds (and you wouldn't want to be under one when it happens). Here, the seed germinates whilst attached to the tree. It rapidly grows a long, straight spear-like appendage hanging vertically downwards like a sword. When the time is right, the seed is shed, and it drops towards the soft sediment. The spear can be over one metre long, and it penetrates the sediment to a great depth. The seedling is instantly rooted and starts to grow.

When considering the animals that live within the mangroves and also importantly, on the mangroves, we find we have two types of zonation: horizontal and vertical. The horizontal zonation is that of a typical estuary where the species distribution is limited by the salt content of the water.

In the vertical plane, we move from purely terrestrial animals and birds down to entirely marine organisms. In the canopy of the mangrove, we can find animals resembling those found in a rainforest. Monkeys, birds (both large and small), snakes, lizards and a full range of insects all inhabit this area. They all contribute to the marine ecosystem, mostly with falling faeces that adds detritus input to the area.

As we enter the intertidal zone, we can leave these behind and start to look at the marine species present. However, there is one exciting snail that lives here, feeding upon the algae growing on the banks, roots and stems: the mangrove snail. It moves up and down the tree, avoiding being submerged and exposed to predators. However, it is not as fast as the tide. It seems to have a built-in clock that registers at some point before the tide turns, and it starts climbing, giving it a head start over the incoming tide.

The soft banks provide homes for many burrowing creatures, such as fiddler crabs and species of prawn and shrimp, all retreating to safety when the tide goes out. Of course, we find our populations of buried shellfish within the mud, filter-feeding once the tidal flow has covered them. One species of goby is able to exist and thrive when the tide recedes, and that is the mudskipper. Here, its front fins have adapted to become stiff and robust, allowing the fish to walk actively over the mud surface. If danger approaches, its short but thick muscular tail comes into action and flips the fish away from danger incredibly quickly.

This ability to survive out of water is facilitated by having two large bags next to its gills, which it fills with water, keeping them moist and oxygenated; then the fish sips at the water's edge to refill them.

The roots and trunks provide a large surface area for shellfish to colonise, and so the primary animal forms in the mangroves are mussels, oysters, and sea squirts - each occupying its own zone with its own resistance to aerial exposure (if any at all). Of the many mobile species that inhabit the area, we find shrimps and prawns in abundance, feeding on the detritus.

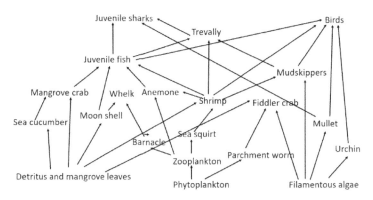

Doodle 25. A typical food web for the mangrove system.

Various crab species are either predators, deposit feeders or filter feeders. All these species act together to provide a great abundance of prey items for the larger predators that enter the mangroves with the incoming tide. Many fish species here are juveniles, using the area as a

111

nursery, as the root systems confer protection from the larger fish. The larger predators are mainly juvenile shark species, along with the odd crocodile. Some of the main predators of the young fish are the numerous species of birds that are present.

Random 6.

Bernadette, the deadly yellow boxfish

Eat or be eaten is the law.

It's a hard life for us under the waves; whether we are planktonic, crawling, sliding, rolling, swimming or buried motionless - we all try to do one thing well and that's eat. We all try to do one thing better, and that's avoid being eaten.

I am a hypocrite. I cannot stand the thought of some big fish with a large mouth grabbing me head-first and sliding me live down its throat to be digested. Yet at the same time, I have no problems biting a happy-go-lucky (but not so fortunate today) shrimp in half. That's the law of the reef for you.

Now I belong to a strange family of fish. Indeed, we are not fast - nope can't outswim you. In fact, we are positively slow. We cannot bury ourselves and we cannot hide in-between small rocks or corals to avoid you. We live mostly solitary lives or indeed in low numbers, so we have no protection in large shoals. We are not camouflaged; in fact, mostly the opposite: we stand out like sore thumbs. Look at 100 fish and who stands out like a lighthouse with a neon sign flashing, 'Here I am! Eat me!' above its head?

So, who am I and why am I still here? Please allow me to introduce myself. I am Bernadette, the sweet, lovely yellow boxfish.

Our little darling babies are the cutest things you have ever seen; they melt even the stoniest hearts. But we pack a dark secret under our skins. One that has taken millions of years to evolve. This excellent defence mechanism really puts the predator off for life - that's if it lives - and to be quite honest, we are rather proud of our evolutionary trick.

Now, I am not going to tell you if we produce our defence with specialised cells under our skins or if it is instead produced by a bacteria living in our skins; that's for you to find out. So what is it? Well, a nasty little chemical called pahutoxin. It packs quite a punch and quickly; well, it has to, or a bite will kill us.

So, let me tell you a tale. I was swimming by a beautiful head of brain coral, when from behind came the jaws of a stonefish; within a second, I was inside and stressed to hell. This stress triggered a fantastic chain reaction, and in milliseconds my toxin was released. Now it's a surfactant, which means it's fast-moving in water and disrupts fats. - Oh, and what are cell membranes made of? Lipids!

Water moves through a fish, and 100% of it flows over the gills, so within 3 milliseconds, a toxin-laden fluid was bathing the predator's breathing apparatus. But I am only a small fish, and the gills of this beast are huge. So it's a good job pahutoxin is lethal in extremely low concentrations - 10 parts per million to be exact. Putting it into human terms: a 5ml-medicine spoonful in an Olympic swimming pool is a deadly mix.

So what does pahutoxin do? It's what you call a haemolytic agent. Haem – blood, lysis – split; so it splits the blood. In fact, it destroys blood cells that carry oxygen into our predator's body. My chemical just sits in the gills, and a biological hammer destroys every blood cell that passes by.

Before the stonefish had a chance to bite down on my fantastic body, it was spitting me out and in a terrible state: jerking and shuddering until it merely died of oxygen starvation. Did it bother me? No. It's the law of life on the reef.

7. CORAL REEFS

For anyone who has the remotest interest in marine life, there are no words to describe the beauty that is the coral reef. It's no wonder that films such as *Avatar* took their inspiration from the life forms in this habitat, which produce such visual amazement. To swim or snorkel over the reef is to be transported to another world, as in the film. Imagine if you were able to hover over the top of a rainforest canopy and remove the leaves, leaving all the animals in full exposure. Then you would begin to find a comparison to the diversity of life that is found here. Not only that, but just like the rainforest, a reef is a noisy place indeed; the sound of the reef coming to life is often likened to a bowl of popping cereal.

The hugely varied and intensely-coloured lizard and birdlife of the rainforest is replaced by the immeasurable array of brightly coloured fish, swimming in and out of the reef. If you take a flashlight out into the rainforest at night, you will be rewarded with an indescribable reflective wall of hundreds of eyes – evidence of the animals lurking in the undergrowth. If you take a blue light down onto the reef at night, the nocturnal coral polyps produce fluorescent light, turning them from dull browns to vivid green, reds and blues; you're dancing in the nightclub that is the coral reef.

Sadly, another comparison we can make to the rainforest is the rate of destruction and damage we find, resulting from human activities. As with the rainforest, there are very few places on earth where we will find a completely undamaged reef system; but more about that later.

So, where do we find coral reefs and what are their limiting factors? You would be forgiven for thinking that coral reefs grow in all tropical regions; after all, they are most commonly associated with tropical islands. In fact, nothing could be further from the truth. Yes, they are only found in the tropics, but due to various biological and physical aspects, they are excluded from many areas there.

What then, is a limiting factor? It is either a biological or physical pressure that stops a plant or animal from growing in a particular area. The most common limiting factor we can think of is temperature. Take a polar bear out of the arctic, and it will soon die of heat; the polar bear species is limited to the arctic by temperature.

So how exactly are the coral reefs limited within the earth's tropical zones? Well, corals are very picky animals indeed; unless they are presented with everything they need, they will not grow. This reflects in their distribution within the tropics.

The first limiting factor is temperature. Ideally corals need an average annual water temperature of between 23 and 25° C and can tolerate slight differences. Below 18° C, we find no coral growth occurs, and with prolonged exposure to 30° C we find coral growth stopping. If the temperature increases any further, the animals become so stressed they will soon die. If we look at the western sides of the major continents such as America and Africa, within the tropical regions, we notice that these are subject to large cold upwellings of water from the deep, resulting in a total absence of corals from these areas.

Now corals don't like to be dirty. What I mean is, they require crystal clear water because they need sunlight to live. They also cannot tolerate sediment being deposited over the top of them. Unless wave action removes it, they have no way of getting rid of this covering, which smothers and kills them. This can result in a total absence of coral from an area. Or if there is a significant disturbance on land, and the river discharges a tremendous amount of sediment, the existing reef there will soon perish. So it is, on the east coast of South America, where conditions are perfect for reef growth. Yet here we have two vast rivers, the Amazon and Orinoco, with their colossal discharge of sediment that causes the exclusion of reefs in an otherwise perfect tropical zone. Localised destruction of reefs through sedimentation is now occurring at an alarmingly regular rate, due to the deforestation of surrounding areas.

Heavy rainfall causes mudslides, then washes sediments into the rivers, smothering the reefs nearby.

Another reason corals will not live near a massive outflow of fresh water is that they are very intolerant to low salinity. However, they can adapt to tolerate higher salinities as in the red sea, where salinity can exceed $40\%_{oo}$.

As previously mentioned, their requirement for light in order to grow and live restricts reef-forming corals to the surface waters. Therefore, not only do we find that corals are limited around the globe, but they are also restricted to the depth of water they can live in. Subsequently, we only really find coral reefs around the margins of the landmasses and islands.

When thinking about the vertical distribution of corals, we must also note that they are extremely intolerant to prolonged aerial exposure and rarely grow in areas that are exposed by the low spring tide. (Please note that we are talking about reef-building corals here, and there are many other species of both reef-forming and individual animals that have adapted to live in deep, dark, cold water.)

We have looked at different types of rocky shores, sandy shores, mangrove thickets and estuaries. Well, coral reefs are no different. We have three generalised reef forms: the fringing reef, the barrier reef and the atoll.

Marine Ecology for the Non-Ecologist

The question we need to answer is: how are they formed? To answer this, we should look at what happened to the earth's crust over geological time. We should also consider the fact that sea levels rose 125 m just over 10,000 years ago with the Ice Age melt.

So let's take a section of the earth's crust - a little piece of rock that is currently residing in the middle of the Pacific Ocean, deep on the seafloor. Now, with the earth's crust moving, our little piece of tectonic plate slowly moves over what is known as a hotspot within the earth itself. This is an area where the magma is closer to the earth's crust than anywhere else, and soon our little piece of rock starts to heat up. As it warms, it becomes less dense and thus begins to rise. Soon the magma of the hotspot pushes through the crust, the top pops off and we have a volcanic eruption under the sea. With each eruption, we find a new layer of rock is added, and over time it builds up and pops out of the ocean surface. We have a new volcanic island; gradually, it grows and grows and grows.

Now around the outskirts of this island, under the sea, we have a new piece of virgin real estate waiting to be colonised. Soon the corals move in, forming a reef around the outskirts of the island. This is our fringing reef, one that completely surrounds the island.

As the earth's tectonic plates are always moving, our volcanic island slowly moves away from the hot spot. It is replaced by a new piece of earth's crust, and a new island grows. Hence, we see chains of islands in the Pacific; one moves away and another is created.

Now, as our volcanic island moves away from the hot spot, the earth's crust is colder, and two things happen. The first is that our volcanic activity simply switches off; the second is that the island starts to sink into the deeper water. As it sinks deeper and deeper, the edge of the reef grows upwards at an equal rate. However, the side of the reef is now further away from the island than it was before. Here we have a barrier reef.

Other barrier reefs are associated with continental landmasses, not islands. Of these, the most famous are located off Belize, in South America, and the Australian West Coast. They are relatively new barrier reefs and were formed with the Ice Age melt. They are found at edge of the continental shelf, which falls steeply into the deep abyss. 10,000 years ago, this area was dry land; the seas rose and washed away the sediment to expose bare rock, and - you guessed it - the corals moved in. As the corals grew with rising sea levels, the majority of the land beside them was protected by the new reef. So the sediment did not wash away, and it stopped any corals from settling and forming substantial reefs. From the reef edge, we have a massive lagoon, sometimes kilometres from the shore.

So the barrier reef in these locations was formed in relatively recent times. In fact, the Australian Aborigines, whose history is recorded in verbal tales and rock art, still talk of the time when their ancestors were hunting where the reef exists today.

We have a situation where not all barrier reefs were once fringing reefs, but all fringing reefs will become barrier reefs.

Back to our little island, which has sunk lower into the sea and is happily existing, protected by its barrier reef. Soon the earth's crust moves further away and into deeper water. Sadly, we wave goodbye to our island, one that we have followed for millions of years, as it sinks beneath the surface. All that is left is the surrounding barrier reef.

Now the soil that covered the island is mixing with the sand that the reef continually creates. Due to tidal currents and wave action, this sand accumulates over different areas of the barrier reef. This causes some parts to die off, whilst other parts continue to thrive. Those areas where the sand has accumulated become more significant and profound. Purely by chance, seeds find their way onto this new land, including coconuts that could have drifted thousands of miles over the ocean surface. They germinate, and new growth occurs, binding the sediments together with its roots, until we have new islands being formed. These islands create a circular shape around the original old, sunken volcanic island that has long since disappeared beneath the waves. Here we have the coral reef system called the atoll.

Fringing Reef   Barrier Reef   Atoll Reef System

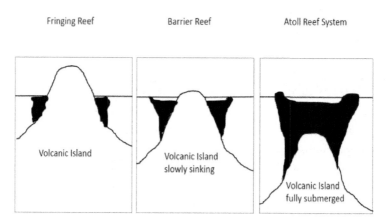

Doodle 26. The evolution of the reef system.

We find many atoll systems in areas such as the Maldives and in the Indo-Pacific regions. However, we do have one more type of coral island that is more commonly found in Indonesia. This is where, in the distant past, the earth's crust moved up only a few feet, but it was enough to push the tops of the corals out of the water. These subsequently died but provided a limestone base for terrestrial plants to colonise; the islands themselves are now surrounded by reefs.

If ever you find yourself on one of these remote islands, you have to be very careful when walking through the interior, as there exist what are known as coral holes. These are ancient caves within the old coral system, and you could soon fall through them. However, I digress; back to the coral reefs themselves.

Within reef systems, we find two distinct separations between the fringing and barrier reefs, which have roughly the same structure as the atoll. Within the different reef zones that are about to be described, we will see that there are totally different environments due to the energy in the area, and light penetration. This in turn will lead to various animal species inhabiting different regions.

Firstly, we shall look at the fringing and barrier reefs. Starting on the island side, we enter the lagoon. This whole area is protected by the reef itself. Thus, this a relatively low-energy environment regarding wave movement. However, depending on its size, we can indeed find powerful currents within the lagoon. The lagoon is characterised by vast expanses of coral sand and coral rubble - broken off bits of coral that have washed over from the front of the reef. Depending on where you are in the world, the lagoon can be less than a metre to many kilometres in length.

Dotted in and around the lagoon are patch reefs, which can vary in size from one coral outcrop to a healthy stand of coral. These are always surrounded by areas of sediment. As we move to the outer edge of the lagoon, we come to the reef flat and back reef. Here the substrate is much larger in size and is made up entirely of broken bits of coral that have been washed over the algal ridge. The back reef experiences a higher wave action due to the waves breaking over the algal ridge. The size distribution of the coral pieces here can range from the size of your thumb to large chunks the size of your leg. All these pieces interlock to form a structure that is very rigid, yet allows water to flow through it; a honeycomb for mobile animals to happily exploit. So here we have another fantastic surface for colonisation. In many cases, the back reef and reef flat can be colonised by various stony corals, which become exposed at low tide. They are able to survive in this area, and indeed thrive, because of the action of the crashing waves over the algal ridge, which continuously sprays the exposed coral with life-giving water.

As we move up the back reef, we encounter the algal ridge - so named because it is made entirely of calcareous algae. It is often exposed to air and takes the full battering of the crashing waves. Nothing can take hold here. It is totally devoid of life apart from the algae from which it takes its name.

Immediately below the algal ridge, we have an expanse known as the buttress zone. This is the area that takes the full force of the tidal action. On this steep slope of around 10m, we find great channels that are carved out of the reef to funnel away vast volumes of high-velocity water, created when the waves are forced into the reef. These are called the surge channels. It was measured that at any one time, the power of this downward surge could be as much as 500,000 horsepower. Here we have large volumes of water, moving away and rushing down the reef until their energy dissipates, then being pushed up with an incoming surge. This is one of the least-studied areas of any reef system, merely due to the fact that if we placed any form of equipment into these channels, it would be destroyed within an instant.

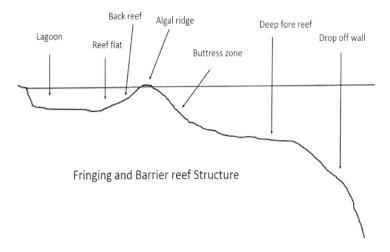

Doodle 27. A typical fringing and barrier reef structure. As always, localised variations do exist.

As the buttress zone levels out, we come to the fore reef and the deep fore reef. A high-energy environment in regard to currents, this is the area we think of when we imagine the coral reef and its life forms. Immense 3D structures extend upwards with overhangs interspersed with channels. Here we find the end of the surge channels depositing the surface water. Deep, dark caves, ledges and hide holes are all in abundance.

After the deep fore reef, we find the drop off wall and, swimming over that, we see the sea descending into blackness. Here we have less energy, a reduction in light levels and dropping temperatures - factors that the animals inhabiting the upper regions of the drop off wall must contend with.

Atoll reefs are both similar yet very different in their construction. To describe them, we take a line through the whole atoll and consider the structure of the reefs in relation to the prevailing wind direction. Remember that wind is the dominant force in surface wave generation. Therefore, one side of the atoll will receive the full force of the waves and tide, whereas the other side will be a relatively low-energy environment regarding wave movement (but not necessarily currents). Here we have two distinct sides: the windward reef and the leeward reef.

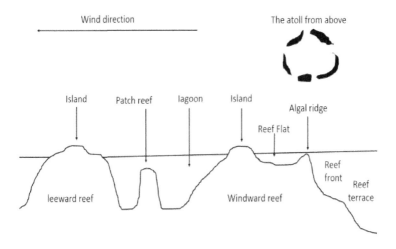

Doodle 28. A cross-section of the typical atoll, showing how it differs with the different physical pressures placed upon it by the prevailing wind.

When we consider the windward reef, we have precisely the same structure as the fringing reef, because it experiences the same environmental conditions. Moving from the algal ridge, we find an extensive reef flat, which terminates at the edge of the windward island.

Behind the island we enter a considerably deep lagoon, often over 100 m in depth. Here it is quite common to find a narrow band of coral-sand beach, which soon drops off into the depths.

Moving across the lagoon, we find that it is interspersed with rock pinnacles, extending from the deep up into the surface waters. These are areas where coral growth has kept up with the sinking island. If you fly over an atoll, you will see quite extensive isolated reefs, surrounded by the dark blue waters. These are the patch reefs. Their size can be as little as 10m across to hundreds of metres, all teeming with life.

Then we move to the leeward island, characterized by a relatively short expanse of coral sand on the lagoon side. Travelling to the leeward reef, we find there is a total absence of the algal ridge and reef flat; there are also no surge channels and buttress zone because of the vastly reduced wave action. This is a result of the protection provided by the other side of the atoll.

Here the reef simply slides away in its beautiful 3D-structure to the drop off wall. The zonation of the corals is far more relaxed, as they gently mix with each other with depth.

There is, however, a zonation that exists within the corals starting at the buttress zone and reef front. Here, healthy and fast-growing corals extend outwards, forming immense finger-like structures called spurs. The spurs grow outwards and across to the surge channels. Sometimes growth is so vigorous that two spurs from either side of a surge channel will meet over the surface, forming a tunnel. As bits of coral are broken off, they are forced up and down the surge channel, further scouring its surface and increasing its depth. It is around this area that we find the highest coral growth: typically between 15 and 20 m depth.

As we reach the reef terrace, we find that the stony coral is less dominant, allowing other sessile animals to either take over entirely or be interspersed between stands of hard corals. The hard corals living here have taken completely different shapes, moving from branched forms to plate-like forms. Commonly the same species of coral can produce either shape. The plate-like style has evolved to allow the coral to capture more light in the deeper areas. We find soft coral, sponges, anemones and a multitude of bivalve species, including the great giant clam. Further down, we find the total absence of any animal that requires light energy to live; here we enter the realm of the filter feeders, with sea fans and sponges dominating this area.

The actual quantity of mobile life on the reef is uncountable and because of its mobility, zonation does not exist. What we do find, especially with fish species, is that juveniles often live their young lives hiding between the coral fingers, which confer protection from anything that would eat them. Shrimps, crabs, starfish, shellfish and molluscs are all in abundance.

In the lagoon we find different species, due to the low-energy and high sediment environment. Here we see an abundance of burrowing species living within the sediments: from shellfish to shrimps, from starfish to sea urchins.

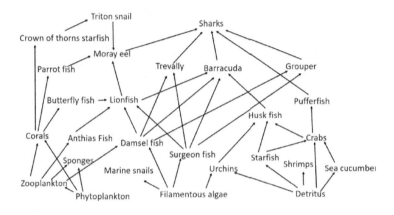

Doodle 29. A generalised food web for a typical coral reef.

There are two significant distinctions between coral reefs located on different sides of the world - that is the Indo-Pacific and the Caribbean. You would be forgiven for thinking that the highest diversity of coral species would be found in the Great Barrier Reef in Australia. The Australian tourist board won't like me for this, but the highest diversity is located within a region called the Coral Triangle in Indonesia. So what we find is that the highest diversity of coral species is in the Indo-Pacific region, whereas the lowest diversity of coral species is in the Caribbean and Hawaii.

In the Indo-Pacific we have reefs that are generally highly diverse, with many species existing side by side; the dominant life form being corals themselves. In the Caribbean, we find vast expanses of coral reef with just one species evident. However, the dominant life forms around the Caribbean are not corals but are, in fact, large tubular sponges.

Now we will broach the subject of reef damage; The main predator that causes extensive destruction is the crown-of-thorns starfish. This beast is raising its head more and more often, appearing at times in herds that clear vast swathes over a reef, leaving behind just the bare calcium rock produced by the corals. Whether this is a natural cycle or their populations are greater, we do not yet know; but when they appear, the damage is vast. From a broad ecological viewpoint, this is a disturbance in the natural progression. From an economic perspective, this is a complete disaster, as no tourists will fly to a place and spend their dollars in an area with no coral.

Back to ecology: what is left is another valuable area of real estate – yes, you guessed it – a new surface to colonise. Human damage to the reef (a subject we will discuss shortly) will have the same effect.

So what will happen? Will new coral larvae settle and reform the reef? Will it be the same coral species as before or different species? Will large algae move into the area, thus changing the entire habitat from coral-based to algae-based, with its associated animal species? The answer is: no one knows until it happens, but as ever with ecology, life will go on!

Human exploitation of the reefs on a global scale is massive; the different forms of destruction are vast. Removal of animals for the aquarium trade, dynamite fishing, mining of reefs for building material, overfishing, mining essential elements for battery-powered cars, pharmaceutical exploitation - the list is endless. There are hardly any reefs left in the world that have not been damaged.

Global warming is of enormous concern for the future of our reefs on a few levels. We have seen a localised temperature rise over reefs, which has caused corals to become stressed over vast areas. This leads to a phenomenon known as bleaching, which can result in coral death. Also, with the increased carbon dioxide in the air being absorbed by the sea, oceans are acidifying, again leading to coral destruction. Sea levels are going to rise; that is a fact. How much they will increase, we do not know. Low-lying islands like atolls will simply be submerged. However, one thing is certain: corals will be able to cope with the sea-level rise as they did 10,000 years ago.

Countless people are working on reducing global warming and its effects. Additionally, efforts to protect coral reefs from further destruction and pollution are being made on a worldwide scale.

I would like to end on a positive note. If we leave things alone - not just coral reefs but all habitats, whether they be marine or terrestrial, they will heal. Hundreds of studies have shown that if we leave the habitat alone, the animals that live there will recover. Look at the great whales that were hunted nearly to extinction but have now been protected by most countries for many years. When we look at their populations, we see that most are currently not endangered but, in fact, quite healthy. Nevertheless, they still need protection from exploitation.

So quite simply: leave things alone, and they will repair.

Random 7.

Sid the Sponge

I am described by you humans as one of the simplest and most primitive multicellular life forms on our earth today. That really ticks me off. How insulting can you get? But then, what else can we expect from a race who think they are above all else? Yes, mate, I have got a chip on my shoulder. However, I can grow my chip back. Could you grow another arm if I chopped it off? So being simple and primitive isn't that bad, is it?!

I am not simple; I am just uncomplicated in my body structure. Actually, I am quite amazing. Please allow me to introduce myself. My own species you humans have not discovered yet, but you know of over 5,000 others. I am a sponge.

I started life as a member of the plankton family, drifting with the currents. At one time, I had a beating whip on my body called a flagellum, which helped me float in the water. I drifted for quite a distance before I landed. After testing the water flow and chemical cues of the rock I was on, I finally settled. Then I 'simply' turned myself inside out. My outer cells with the flagella migrated inwards, and my inner cells migrated outwards - a funny feeling I can say!

That was a long time ago. Now, I am a big boy, and lining my exterior are millions of cells called porocytes. Each has a tunnel running right through it, allowing the passage of water. The pores can close at will, turning the water supply on and off. The water passes through loads of tunnels, which eventually meet in a large central exhalent tube.

Along the tunnel walls live millions of cells, each with a beating flagellum. These are named collar cells because of the collar of microvilli around the base of each flagellum. The beating action draws water in from the sea, then microscopic bits of detritus and dissolved organics are absorbed into the cells by the microvilli - food, lovely food.

As water passes the cell walls, waste from my body is excreted from them and carried by the water into the exhalent tunnel. Here it passes upwards to the big hole in the top called the osculum. This big exhalent opening is assisted by the design we sponges created; indeed, we are quite proud of this feature. The shape of the osculum causes a localised drop in water pressure around it. As a result, the exhalent water is sucked out of our bodies, up and away, to stop us re-ingesting it. I am 10 cm tall and filter 100 litres of water a day. I must congratulate myself, 'not bad old chap, not bad at all'.

My body structure is supported by large silica spines called spicules. My mates make theirs out of calcium or sponging, which is a soft organic material. When you lot cut us up, this is what is used to identify the species. We are then mashed to a pulp so you can discover which chemicals we produce. Many of our chemicals have helped in the treatment of various diseases, even tumours. Someone once pulped one of us and then sieved the little particles of solids that were left; they placed these in an experimental aquarium, and within months my old friend was growing all over the place!

Marine Ecology for the Non-Ecologist

I will have to go now as I can sense a large concentration of dissolved organics coming my way - dinner time.

Index

Random 8.

Sir Isaac Newton, William Herschel, light and fluorescent corals

Just recently, I have seen a few articles on night diving or fluorescent coral diving. I really don't know if this is a new area to the SCUBA lifestyle, but I am certain that any diver willing to take the night dive will, quite simply, experience the dive of their life. It's fantastic to think that animals as small as corals can provide a 'wow factor' equal to or exceeding that of diving with manta rays, whale sharks, or any other beast residing in our fantastic oceans.

Any divers taking a blue light down onto a reef at night will be rewarded with corals fluorescing back at them in the most fantastic fashion; the polyps glow green, red, blue and yellow. It is truly the eighth wonder of the natural world.

So what has William Herschel - the great man who discovered the planet Uranus - got to do with this? Well, it's all related to the area of the electromagnetic spectrum called visible light and an experiment he did in the year 1800.

Nearly 100 years earlier (in 1704), another great, Mr. Newton, first described light as being made up of many colours by passing it through a prism. Herschel then placed thermometers into each colour when he passed light through a prism; each colour had a different temperature. Or think of it like this: a different energy level.

Why did they each have a different energy level? Well, simply put, each colour has a different wavelength and frequency. Frequency is the number of waves passing a point in one second; the higher the rate, the more waves there are and the higher the energy level. So when the light exits the prism, it splits into its different wavelengths, and that's what we see: energy at different wavelengths.

So our coral polyps absorb light at specific wavelengths, then re-emit it. In the process, some energy is lost as heat. The outgoing light has less energy, and thus a longer wavelength than the light being absorbed. The change in wavelength means a shift in colour; this is fluorescence - and what a sight it is!

Evolution is a funny thing. You might wonder why this has occurred. Well, I would like to tell you, but the fact is, we don't really know. As with most things, I don't think it will be for only one reason, but more likely an interplay of factors providing the corals with a number of advantages.

Perhaps the heat loss could aid healing processes in a damaged polyp. Maybe its tissues absorb ultraviolet light and re-emit it as a natural sunscreen to protect the animal. Or, importantly, could the symbiotic algae contained within the polyp be a prey-capture-system that attracts animals to the polyps' stinging tentacles?

I love it when physics and biology come together, as they always do. Fantastic, isn't it? You only have to take a look at the fluorescent corals to be amazed at their beauty; truly the eighth wonder of the biological - or should I say, the *natural* world.

Marine Life
A three-book series by
Andrew Caine

**<u>Marine Biology
for the
Non-Biologist</u>**

THE BASIS OF LIFE IN THE OCEANS

SOFT-BODIED ANIMALS - THE CNIDARIANS
THE JELLYFISH
THE HYDROIDS
THE ANEMONES
THE CORALS
THE STONY CORALS

SHELLFISH - THE MOLLUSCS

THE GASTROPODS
THE MESOGASTROPODS
THE ARCHEOGASTROPODS
THE NEOGASTROPODS
THE NUDIBRANCHS (SEA SLUGS)
THE BIVALVES
THE CEPHALOPODS

THE NAUTILUS
THE SQUIDS, CUTTLEFISH AND OCTOPI

ANIMALS WITH EXOSKELETONS - THE CRUSTACEANS

THE DECAPODS- CRABS, LOBSTERS AND SHRIMPS
THE BARNACLES

ANIMALS WITH SPINY SKINS - THE ECHINODERMS

THE STARFISH
THE BRITTLE STARS
THE SEA URCHINS
THE SEA CUCUMBERS
CORAL REEF ARCHITECTURE

MARINE INVERTEBRATE TOXINS

LIMU-MAKE-O-HANA (THE DEADLY SEAWEED OF HANA)
SHELLFISH POISONING
PARALYTIC SHELLFISH POISONING (PSP)
NEUROTOXIC SHELLFISH POISONING (NSP)
DIARRHOEIC SHELLFISH POISONING (DSP)
AMNESIC SHELLFISH POISONING (ASP)
CIGUATERA
THE CONE SHELLS
SEA SNAKES
VENOMOUS FISH

HYDROTHERMAL VENTS AND VENT BIOLOGY

THE DISCOVERY
THE PHYSICAL ENVIRONMENT
VENT BIOLOGY
LIFE IN THE POLAR SEAS

THE POLAR ENVIRONMENT

ANIMAL ADAPTATIONS
THE FISH
THE INVERTEBRATES
THE BIRDS
THE MAMMALS

<u>Incredible Oceans</u>

Amazing facts and explanations from the wonderful worlds of:

Marine Biology
Marine Ecology
and
Oceanography

THE OCEAN

IT'S ONE BODY OF WATER AROUND
THE GLOBE AND IT'S AMAZING.

THE PLANKTON

YOU CAN'T SEE MOST OF THEM.
JUST WAIT UNTIL YOU HEAR ABOUT THEM;
LIFE-GIVING YET DEADLY AND SILENT!

LOCOMOTION AND MIGRATIONS

FROM THE SLOW TO THE SUPER SPEED,
IT'S ALL A MATTER OF LIFE AND DEATH
GOING ON HOLIDAY AND WHY! NO
WAITING IN AN AIRPORT AND NO
SAT NAV EITHER.

FEEDING

FINE DINING TO FLUID ONLY,
NO TABLE MANNERS HERE

REPRODUCTION AND LIFE SPANS

PASSING ON THAT DNA, OR JUST MAKING
A COPY, MIND-BLOWING ACTIVITIES AND
STRATEGIES.
LIFE FROM HOURS TO 100S OF YEARS, WHO
WAS SWIMMING WHEN QUEEN ELIZABETH 1
WAS ON THE THRONE.

HOUSING

ONE OF THE BIGGEST SHORTAGES TO HIT
THE OCEAN LIFE, REAL ESTATE. WE ANSWER
THE IMMORTAL QUESTION –
WHO LIVES IN A HOUSE LIKE THIS?

RELATIONSHIPS BETWEEN SPECIES

WHO HELPS WHO AND WHY, OR PARASITES?
TO SUCK YOU DRY?

WHATS GOING TO KILL YOU?

YOU DON'T WANT TO MEET THESE - EATING
ONE IS DEADLY TOO.

POLLUTION AND DESTRUCTION

THE SOLUTION TO POLLUTION IS DILUTION.
A SAYING ORIGINATING FROM THE 1960s.
ONLY NOW THE OCEAN IS FULL UP!

Printed in Great Britain
by Amazon